Intermediate Rigger

Trainee Guide

Boston   Columbus   Indianapolis   New York   San Francisco   Amsterdam
Cape Town Dubai London Madrid Milan Munich Paris Montreal Toronto Delhi
Mexico City   Sao Paulo   Sydney   Hong Kong   Seoul   Singapore   Taipei   Tokyo

NCCER

President: Don Whyte
Vice President: Steve Greene
Chief Operations Officer: Katrina Kersch
Rigger Curriculum Project Manager: Chris Wilson
Senior Development Manager: Mark Thomas

Senior Production Manager: Tim Davis
Quality Assurance Coordinator: Karyn Payne
Desktop Publishing Coordinator: James McKay
Permissions Specialists: Kelly Sadler
Production Specialist: Kelly Sadler
Editors: Graham Hack, Debie Hicks

Writing and development services provided by Topaz Publications, Liverpool, NY

Lead Writer/Project Manager: Troy Staton
Desktop Publisher: Joanne Hart
Art Director: Alison Richmond

Permissions Editor: Andrea LaBarge
Writers: Troy Staton, Thomas Burke, Terry Egolf

Pearson

Director of Alliance/Partnership Management: Andrew Taylor
Editorial Assistant: Collin LaMothe
Program Manager: Alexandrina B. Wolf
Assistant Content Producer: Alma Dabral
Digital Content Producer: Jose Carchi
Director of Marketing: Leigh Ann Simms

Senior Marketing Manager: Brian Hoehl
Composition: NCCER
Printer/Binder: LSC Communications
Cover Printer: LSC Communications
Text Fonts: Palatino and Univers

Credits and acknowledgments for content borrowed from other sources and reproduced, with permission, in this textbook appear at the end of each module.

ISBN-13:   978-0-13-518320-5
ISBN-10:   0-13-518320-0

ScoutAutomatedPrintCode

To the Trainee

Rigging is an activity performed on virtually every construction site and in every industrial facility. Becoming a trained rigger will open up the doors of opportunity and you will be able to choose to work in construction, the power industry, the petroleum industry, the maritime industry, the mining industry, or the manufacturing industry.

As you progress through your training, you will be advancing your knowledge and skills with progressively challenging topics and activities. Riggers who continue to advance their knowledge have plenty of room for job growth. After gaining experience, advanced riggers can become rigging foremen or lift directors.

The need for riggers will continue to increase in all industries as experienced riggers retire. Those who meet the qualifications under OSHA 29 *CFR* Part 1926 Subpart CC, *Cranes and Derricks in Construction* will continue to be in high demand in this field that the US Department of Labor expects to experience faster than average growth over the next several years.

New with *Intermediate Rigger*

NCCER is pleased to release *Intermediate Rigger* in full color, with new photographs and figures. This edition is now presented in NCCER's improved instructional systems design, in which the sections of each module are directly tied to learning objectives.

The suggested training time for this Intermediate Rigger program has nearly doubled because the training is now more thorough. The "Intermediate Rigging" module features new training on how to turn and invert a hoisted load and how to drift a load from one hoist to another. The "Load Dynamics" module expands the training on the physics behind lifting and moving loads. The modules on telescopic boom attachments and lattice boom assembly/disassembly also provides expanded information for riggers who work on and around mobile cranes.

We wish you success as you progress through this training program. If you have any comments on how NCCER might improve upon this textbook, please complete the User Update form located at the back of each module and send it to us. We will always consider and respond to input from our customers.

We invite you to visit the NCCER website at **www.nccer.org** for information on the latest product releases and training, as well as online versions of the *Cornerstone* magazine and Pearson's NCCER product catalog.

Your feedback is welcome. You may email your comments to **curriculum@nccer.org** or send general comments and inquiries to **info@nccer.org**.

NCCER Standardized Curricula

NCCER is a not-for-profit 501(c)(3) education foundation established in 1996 by the world's largest and most progressive construction companies and national construction associations. It was founded to address the severe workforce shortage facing the industry and to develop a standardized training process and curricula. Today, NCCER is supported by hundreds of leading construction and maintenance companies, manufacturers, and national associations. The NCCER Standardized Curricula was developed by NCCER in partnership with Pearson, the world's largest educational publisher.

Some features of the NCCER Standardized Curricula are as follows:

- An industry-proven record of success
- Curricula developed by the industry, for the industry
- National standardization providing portability of learned job skills and educational credits
- Compliance with the Office of Apprenticeship requirements for related classroom training (*CFR* 29:29)
- Well-illustrated, up-to-date, and practical information

NCCER also maintains the NCCER Registry, which provides transcripts, certificates, and wallet cards to individuals who have successfully completed a level of training within a craft in NCCER's Curricula. *Training programs must be delivered by an NCCER Accredited Training Sponsor in order to receive these credentials.*

Special Features

In an effort to provide a comprehensive and user-friendly training resource, this curriculum showcases several informative features. Whether you are a visual or hands-on learner, these features are intended to enhance your knowledge of the construction industry as you progress in your training. Some of the features you may find in the curriculum are explained below.

Introduction

This introductory page, found at the beginning of each module, lists the module Objectives, Performance Tasks, and Trade Terms. The Objectives list the knowledge you will acquire after successfully completing the module. The Performance Tasks give you an opportunity to apply your knowledge to real-world tasks. The Trade Terms are industry-specific vocabulary that you will learn as you study this module.

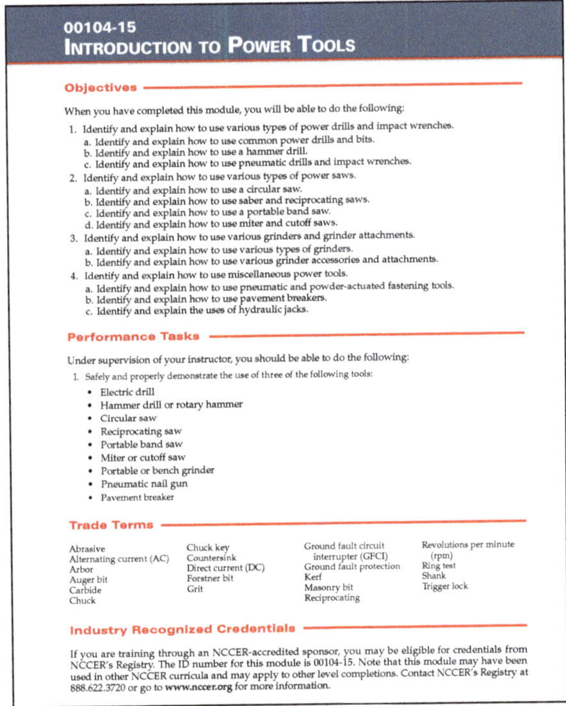

Trade Features

Trade features present technical tips and professional practices based on real-life scenarios similar to those you might encounter on the job site.

Bowline Trivia

Some people use this saying to help them remember how to tie a bowline: "The rabbit comes out of his hole, around a tree, and back into the hole."

Figures and Tables

Photographs, drawings, diagrams, and tables are used throughout each module to illustrate important concepts and provide clarity for complex instructions. Text references to figures and tables are emphasized with *italic* type.

Notes, Cautions, and Warnings

Safety features are set off from the main text in highlighted boxes and categorized according to the potential danger involved. Notes simply provide additional information. Cautions flag a hazardous issue that could cause damage to materials or equipment. Warnings stress a potentially dangerous situation that could result in injury or death to workers.

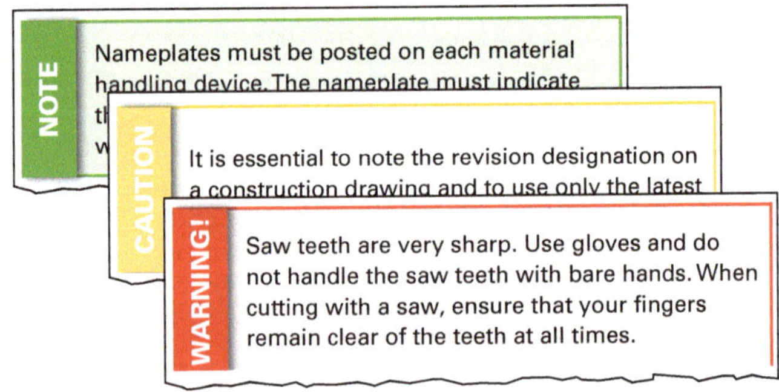

Case History

Case History features emphasize the importance of safety by citing examples of the costly (and often devastating) consequences of ignoring best practices or OSHA regulations.

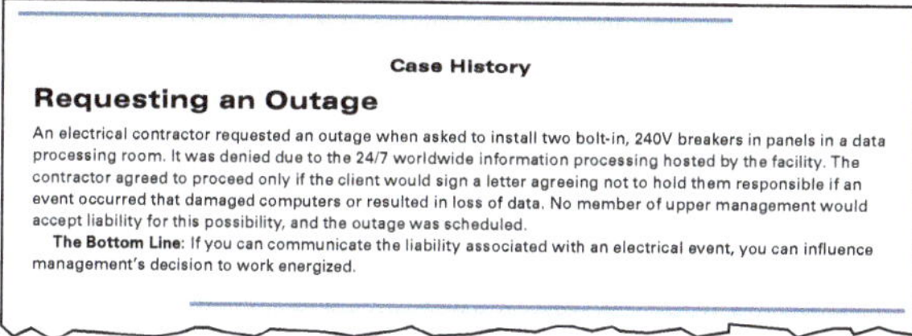

Going Green

Going Green features present steps being taken within the construction industry to protect the environment and save energy, emphasizing choices that can be made on the job to preserve the health of the planet.

Did You Know

Did You Know features introduce historical tidbits or interesting and sometimes surprising facts about the trade.

Step-by-Step Instructions

Step-by-step instructions are used throughout to guide you through technical procedures and tasks from start to finish. These steps show you how to perform a task safely and efficiently.

Perform the following steps to erect this system area scaffold:

Step 1 Gather and inspect all scaffold equipment for the scaffold arrangement.

Step 2 Place appropriate mudsills in their approximate locations.

Step 3 Attach the screw jacks to the mudsills.

Trade Terms

Each module presents a list of Trade Terms that are discussed within the text and defined in the Glossary at the end of the module. These terms are presented in the text with bold, blue type upon their first occurrence. To make searches for key information easier, a comprehensive Glossary of Trade Terms from all modules is located at the back of this book.

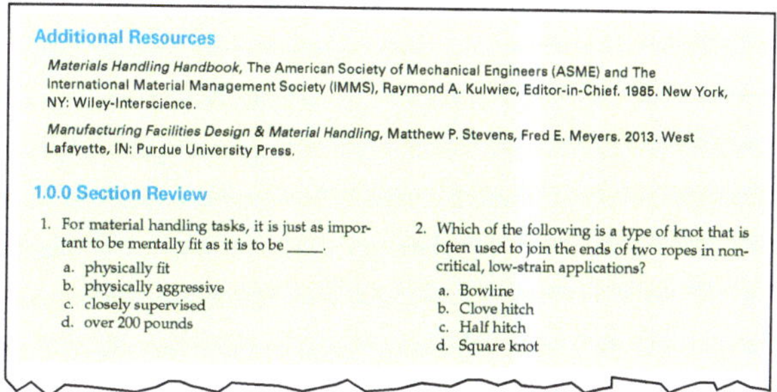

During a rigging operation, the load being lifted or moved must be connected to the apparatus, such as a crane, that will provide the power for movement. The connector—the link between the load and the apparatus—is often a sling made of synthetic, chain, or wire rope materials. This section focuses on three types of slings:

Section Review

Each section of the module wraps up with a list of Additional Resources for further study and Section Review questions designed to test your knowledge of the Objectives for that section.

Review Questions

The end-of-module Review Questions can be used to measure and reinforce your knowledge of the module's content.

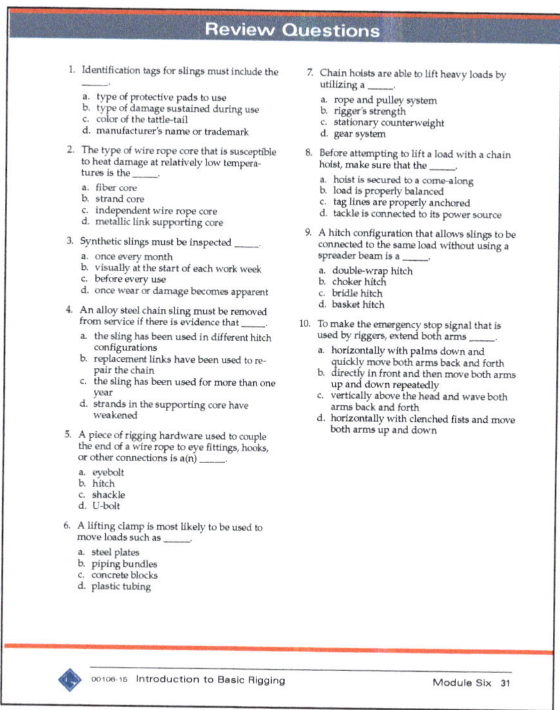

NCCER Standardized Curricula

NCCER's training programs comprise more than 80 construction, maintenance, pipeline, and utility areas and include skills assessments, safety training, and management education.

Boilermaking
Cabinetmaking
Carpentry
Concrete Finishing
Construction Craft Laborer
Construction Technology
Core Curriculum: Introductory
 Craft Skills
Drywall
Electrical
Electronic Systems Technician
Heating, Ventilating, and Air
 Conditioning
Heavy Equipment Operations
Heavy Highway Construction
Hydroblasting
Industrial Coating and Lining
 Application Specialist
Industrial Maintenance Electrical
 and Instrumentation Technician
Industrial Maintenance Mechanic
Instrumentation
Ironworking
Manufactured Construction
 Technology
Masonry
Mechanical Insulating
Millwright
Mobile Crane Operations
Painting
Painting, Industrial
Pipefitting
Pipelayer
Plumbing
Reinforcing Ironwork
Rigging
Scaffolding
Sheet Metal
Signal Person
Site Layout
Sprinkler Fitting
Tower Crane Operator
Welding

Maritime

Maritime Industry Fundamentals
Maritime Pipefitting
Maritime Structural Fitter

Green/Sustainable Construction

Building Auditor
Fundamentals of Weatherization
Introduction to Weatherization
Sustainable Construction
 Supervisor
Weatherization Crew Chief
Weatherization Technician
Your Role in the Green
 Environment

Energy

Alternative Energy
Introduction to the Power Industry
Introduction to Solar Photovoltaics
Power Generation Maintenance
 Electrician
Power Generation I&C
 Maintenance Technician
Power Generation Maintenance
 Mechanic
Power Line Worker
Power Line Worker: Distribution
Power Line Worker: Substation
Power Line Worker: Transmission
Solar Photovoltaic Systems Installer
Wind Energy
Wind Turbine Maintenance
 Technician

Pipeline

Abnormal Operating Conditions,
 Control Center
Abnormal Operating Conditions,
 Field and Gas
Corrosion Control
Electrical and Instrumentation
Field and Control Center
 Operations
Introduction to the Pipeline
 Industry
Maintenance
Mechanical

Safety

Field Safety
Safety Orientation
Safety Technology

Supplemental Titles

Applied Construction Math
Tools for Success

Management

Construction Workforce
 Development Professional
Fundamentals of Crew Leadership
Mentoring for Craft Professionals
Project Management
Project Supervision

Spanish Titles

Acabado de concreto: nivel uno
 (*Concrete Finishing Level One*)
Aislamiento: nivel uno
 (*Insulating Level One*)
Albañilería: nivel uno
 (*Masonry Level One*)
Andamios (*Scaffolding*)
Carpintería: Formas para
 carpintería, nivel tres
 (*Carpentry: Carpentry Forms, Level
 Three*)
Currículo básico: habilidades
 introductorias del oficio
 (*Core Curriculum: Introductory Craft
 Skills*)
Electricidad: nivel uno
 (*Electrical Level One*)
Herrería: nivel uno
 (*Ironworking Level One*)
Herrería de refuerzo: nivel uno
 (*Reinforcing Ironwork Level One*)
Instalación de rociadores: nivel uno
 (*Sprinkler Fitting Level One*)
Instalación de tuberías: nivel uno
 (*Pipefitting Level One*)
Instrumentación: nivel uno, nivel
 dos, nivel tres, nivel cuatro
 (*Instrumentation Levels One through
 Four*)
Orientación de seguridad
 (*Safety Orientation*)
Paneles de yeso: nivel uno
 (*Drywall Level One*)
Seguridad de campo
 (*Field Safety*)

Acknowledgments

This curriculum was revised as a result of the farsightedness and leadership of the following sponsors:

ABC Pelican Chapter
Bay Ltd.
Bechtel
Bo-Mac Contractors, Ltd.
Cowboyscranes.com
Exelon Generation
Fluor Corp.

KBR, Inc.
Kelley Construction
Mammoet USA
North American Crane Bureau
Orion Marine Group
Southland Safety

This curriculum would not exist were it not for the dedication and unselfish energy of those volunteers who served on the Authoring Team. A sincere thanks is extended to the following:

Ed Burke
Robert Capelli
Anthony Johnson
Richard Laird
Steven Lawrence

Don McDonald
Timothy Prakop
Larry "Cowboy" Proemsey
Joseph Watts
Harold Williamson

A sincere thanks is also extended to the dedication and assistance provided by the following technical advisors:

Monty Chisolm   Keith Denham   Frank Jones

NCCER Partners

American Council for Construction Education
American Fire Sprinkler Association
Associated Builders and Contractors, Inc.
Associated General Contractors of America
Association for Career and Technical Education
Association for Skilled and Technical Sciences
Construction Industry Institute
Construction Users Roundtable
Design Build Institute of America
GSSC – Gulf States Shipbuilders Consortium
ISN
Manufacturing Institute
Mason Contractors Association of America
Merit Contractors Association of Canada
NACE International
National Association of Women in Construction
National Insulation Association
National Technical Honor Society
National Utility Contractors Association
NAWIC Education Foundation
North American Crane Bureau
North American Technician Excellence
Pearson

Prov
SkillsUSA®
Steel Erectors Association of America
U.S. Army Corps of Engineers
University of Florida, M. E. Rinker Sr., School of Construction Management
Women Construction Owners & Executives, USA

Contents

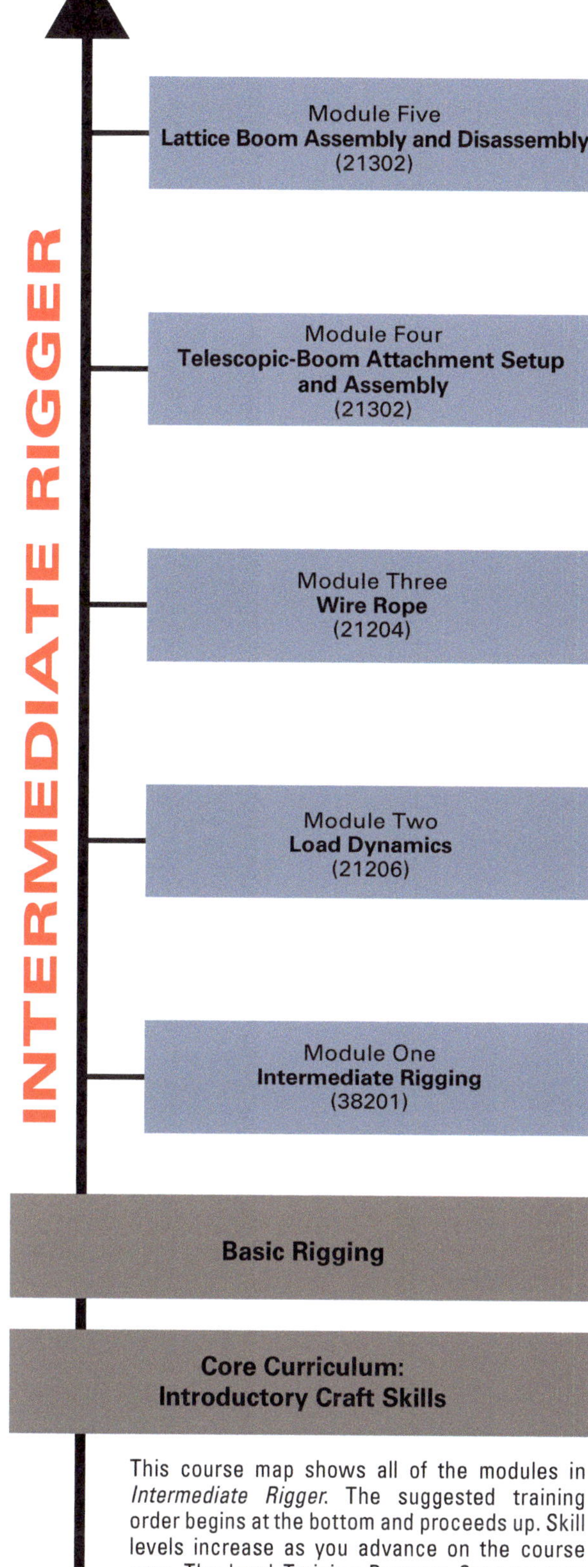

This course map shows all of the modules in *Intermediate Rigger*. The suggested training order begins at the bottom and proceeds up. Skill levels increase as you advance on the course map. The local Training Program Sponsor may adjust the training order.

Intermediate Rigging

OVERVIEW

Riggers need to understand the equipment and procedures used to move loads, both with and without the use of a crane. In many cases, riggers are required to move awkward and heavy loads indoors. This module reviews the selection of slings and hardware used by riggers, and presents some of the equipment used to move loads vertically and laterally. It describes how to turn and invert loads using hoists, and how to move loads over obstacles. The module also explains how to determine the stress placed on slings and anchorage points during load movement.

Module 38201

Trainees with successful module completions may be eligible for credentialing through the NCCER Registry. To learn more, go to **www.nccer.org** or contact us at 1.888.622.3720. Our website has information on the latest product releases and training, as well as online versions of our *Cornerstone* magazine and Pearson's product catalog.

Your feedback is welcome. You may email your comments to **curriculum@nccer.org**, send general comments and inquiries to **info@nccer.org**, or fill in the User Update form at the back of this module.

This information is general in nature and intended for training purposes only. Actual performance of activities described in this manual requires compliance with all applicable operating, service, maintenance, and safety procedures under the direction of qualified personnel. References in this manual to patented or proprietary devices do not constitute a recommendation of their use.

38201 V3

Objective

When you have completed this module, you will be able to do the following:

1. Identify and describe how to use rigging equipment and hoists to move loads.
   a. Identify and describe sling hitches and usage guidelines.
   b. Identify and describe equipment used for lateral load movement.
   c. Describe how to apply and reeve block and tackle.
   d. Describe how to turn and invert a hoisted load.
   e. Describe how to lift and transfer a load from one hoist to another.

Performance Tasks

Under the supervision of your instructor, you should be able to do the following:

1. Reeve block and tackle with proper rope termination.

2. Invert a load using hoists.

3. Drift a load from one hoist to another.

Trade Terms

Anchorage point
Center of gravity (CG)
Choke angle
Coefficient of friction (CF)
D/d ratio
Derate
Drifting

Hitch
Invert
Sling angle
Sling-angle factor (SAF)
Tuggers
Working load limit (WLL)

Industry Recognized Credentials

If you are training through an NCCER-accredited sponsor, you may be eligible for credentials from NCCER's Registry. The ID number for this module is 38201. Note that this module may have been used in other NCCER curricula and may apply to other level completions. Contact NCCER's Registry at 888.622.3720 or go to **www.nccer.org** for more information.

Contents

1.0.0 RIGGING EQUIPMENT AND LOAD MOVEMENT

Objective

Identify and describe how to use rigging equipment and hoists to move loads.

a. Identify and describe sling hitches and usage guidelines.
b. Identify and describe equipment used for lateral load movement.
c. Describe how to properly reeve block and tackle.
d. Describe how to lift and invert a load using hoists.
e. Describe how to lift and transfer a load from one hoist to another.

Performance Tasks

1. Reeve block and tackle with proper rope termination.
2. Invert a load using hoists.
3. Drift a load from one hoist to another.

Trade Terms

Anchorage point: A point used to fix the position of a block or other type of hoist or pulling device. An anchorage point must be able to easily handle the load weights and line pulls associated with any sling angle imposed on it, without fear of failure.

Center of gravity (CG): The point where an object's mass, and therefore its weight, is concentrated. The concept is useful for determining stability, the balance point, and leverage.

Choke angle: The outside angle between the vertical part of a choker hitch and the part of the sling passing through the eye, hook, or shackle forming the choke.

Coefficient of friction (CF): A ratio that expresses a comparison between the force necessary to move an object over the surface of another material, and the pressure between the two materials.

D/d ratio: For rope-like slings used in hitches that wrap around the load, the ratio of the diameter of the load to the diameter of the rope. Manufacturers establish minimum D/d ratios to avoid overstressing the sling with a bend that is too sharp.

Derate: The process of reducing the working load limit of a sling according to standard rules to account for the large stresses that exist in a sling when used in certain applications.

Drifting: The process of moving a load laterally through the air, using multiple hoists and transferring the load weight from one hoist to another.

Hitch: Any method of attaching a sling to a load. Common hitches include vertical, bridle, basket, and choker.

Invert: To move a load from the horizontal position, for example, to any other position, such as over on its side, vertical on its end, or fully inverted to lie on its top.

Sling angle: The angle formed between the sling and the horizontal when under tension.

Sling-angle factor (SAF): A factor that permits calculating the tension in a sling supporting a known load at an angle. Equal to the length of the sling between its bearing points divided by the vertical distance between those points ($SAF = L \div H$).

Tuggers: Motorized or engine-powered winches designed to replace manpower to pull the rope used for lifting or moving equipment.

Working load limit (WLL): The weight capacity a manufacturer certifies a rigging component can lift under stated conditions. The standard sling WLL is provided for a vertical hitch. Riggers will typically apply standard rules to derate WLLs when slings are used in other configurations.

Most lifts involving cranes require some method of attaching the load to the crane's hook. Some form of sling or combination of slings is generally used to accomplish this. Riggers are also responsible for moving loads laterally, and for precise positioning of equipment. These tasks may require the use of special tools and equipment. This module examines the use of various rigging devices and equipment to move loads in a variety of ways.

1.1.0 Slings

The *American Society of Mechanical Engineers* standard for slings, *ASME B30.9, Slings*, defines a sling used for lifting or moving loads as a flexible length of chain, metal mesh, synthetic webbing, or natural, synthetic, or wire rope that connects a load to a hoisting device. For this purpose, a sling may have a loop, or eye, at one or both ends. Instead of these loops, or in addition to them, slings may have hardware for making connections. Both the ASME and OSHA standards provide definitions of the terminology related to hoisting slings.

Riggers use some type of sling in nearly every load handling operation. They must know how to do the following:

- Select the correct sling for each situation.
- Properly use the sling.
- Determine the sling's working load limit (WLL) for a given application.
- Derate the sling's capacity to compensate for various lift factors.

Figure 1 illustrates how slings are measured. In WLL calculations involving slings, a sling's length is considered the distance between the loadbearing points at each end. The widths of the eyes and any permanently attached rings or fittings are measured by the inside dimensions at their widest points.

The following are general requirements for maintaining, selecting, and using slings:

- Do not use damaged or defective slings.
- Do not shorten slings using knots, bolts, or other makeshift devices.
- Do not kink sling legs.
- Do not load slings beyond their rated capacity.
- Balance and secure loads to prevent slippage when using slings in a basket hitch.
- Securely attach slings to their loads.
- Pad or protect slings from sharp edges on their loads.
- Keep fingers and hands clear of loads as slings tighten during a lift.
- Start and stop slowly—do not shock load slings.
- Do not pull slings from under the load when the load is resting on the sling.

- Employers must only use slings with affixed and legible identification markings.
- A competent person must inspect all slings daily prior to use and during use.
- Keep written records of inspections and repairs.

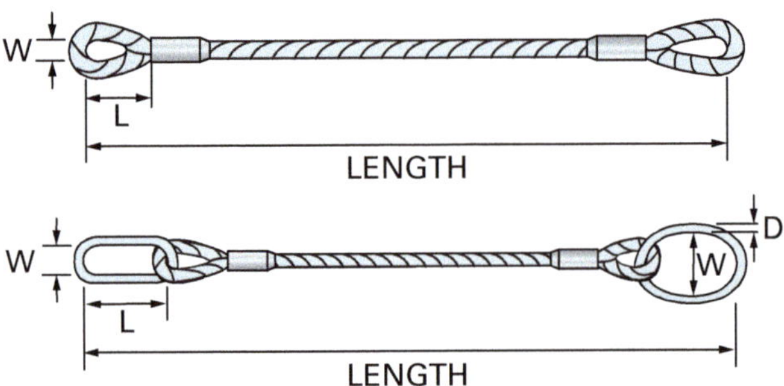

Figure 1 Sling dimensions.

- Observe the operating temperature specified in the standard as well as the manufacturer's guidelines for each type of sling
- Remove slings from service according to the criteria listed in the standard for each type of sling.

ASME B30.9 includes the following recommendations for the proper use and care of slings:

- Store slings in a cool, dry place out of direct sunlight.
- Do not drag slings on the floor or ground.
- Select the type of sling best suited for the load and conditions.
- Make attachments to a load above its center of gravity (CG).
- Use a shackle when using more than one sling at the same point of attachment.
- Lift the load a few inches and then check the rigging and load for stability.
- Select a hitch that will hold and control the load.
- Maintain an appropriate load-diameter to sling-diameter ratio (D/d ratio).
- Secure all unused sling legs when using more than one on a hook.
- Calculate the tension of each sling if the distribution of weight is in question.
- Calculate the vertical hitch capacity reductions when using choker hitches.
- Determine the choke angle to identify any sling capacity reduction when seating or cinching a choker hitch.
- Do not use wire rope clips to form eyes in slings for lifting purposes.
- Always wear gloves when handling slings.

1.1.1 Sling Tension

A vertical hitch is a single sling attached to the load by its end fittings, in which the sling is within 5 degrees of being vertical. Riggers should strive to attach a vertical hitch to a point directly above the load's center of gravity so that the load is naturally balanced.

Multiple vertical hitches can support a load if they are attached to a spreader bar or equalizer beam. The sling's tension in a vertical hitch is equal to the portion of the load that it bears. Therefore, the sling load must be less than its WLL. However, when using a sling in other types of hitches, the sling may support the load at an angle. This increases the tension in the sling compared to its load weight, reducing its load-lifting capacity.

The sling angle is the angle formed by a sling measured from the horizontal plane when under tension. When using sling angle charts (*Figure 2*), remember that it is the horizontal plane, not the vertical plane, that is the reference for stating the sling angle. Because the sling angle directly affects the WLL of any sling, larger sling angles are best for safe rigging.

Assume that two uniformly loaded slings lift a 2,000-pound object, as shown in *Figure 2*. When lifting vertically (vertical hitches), each will carry 1,000 pounds—the sling load. The sling angle in this case is 90 degrees. As the sling angle decreases, each sling is now pulling toward the middle of the load as well as upward. The tension in the sling then includes the 1,000 pounds it is lifting, plus the force pulling toward the middle. As the sling angle decreases, the pull towards the middle increases, resulting in additional tension.

To calculate the total tension in a sling due to its sling angle, first determine its sling-angle factor (SAF), then multiply the sling load by the SAF. Calculate the SAF by dividing the bearing-to-bearing length of the sling (L) by the height difference between its lower and upper bearing points (H), as shown in *Figure 3*.

Riggers should try to keep sling angles greater than 45 degrees. Optimum sling angles fall between 60 and 45 degrees when sling-angle factors are close to 1. Use sling angles between 30 and 45 degrees with caution. When sling angles less than 30 degrees must be used, a qualified person (as defined by OSHA) must evaluate the capacity and condition of the rigging involved. In most cases, there is a better, safer approach that reduces sling tension.

Manufacturers provide sling capacity charts showing the load capacities of their products in various configurations. Such charts are an effective tool for identifying a sling with the necessary WLL for the intended application. Factors to consider include the sling's material, construction, condition, size, hitch configuration, quantity, and intended sling angle. Keep in mind that chart data assumes a sling in new condition.

Sling capacity charts are for general reference when selecting slings for a lift. You can calculate the exact capacity of a sling by determining the sling's material and classification, size, hitch configuration, and sling angle. Riggers are not expected to memorize long tables of working load limits for various sling configurations.

One valuable resource for this information is an aid provided by The Crosby Group. Entitled *Crosby® Users Guide for Lifting*, this aid has long been available as a pocket-sized folding reference that contains a great deal of information for lifting applications (*Figure 4*). Be sure to familiarize yourself with this aid, as it is often one of the few resources that may be used during certification testing. It contains two charts frequently used for reference, Wire Rope Sling Capacities (*Figure 5*)

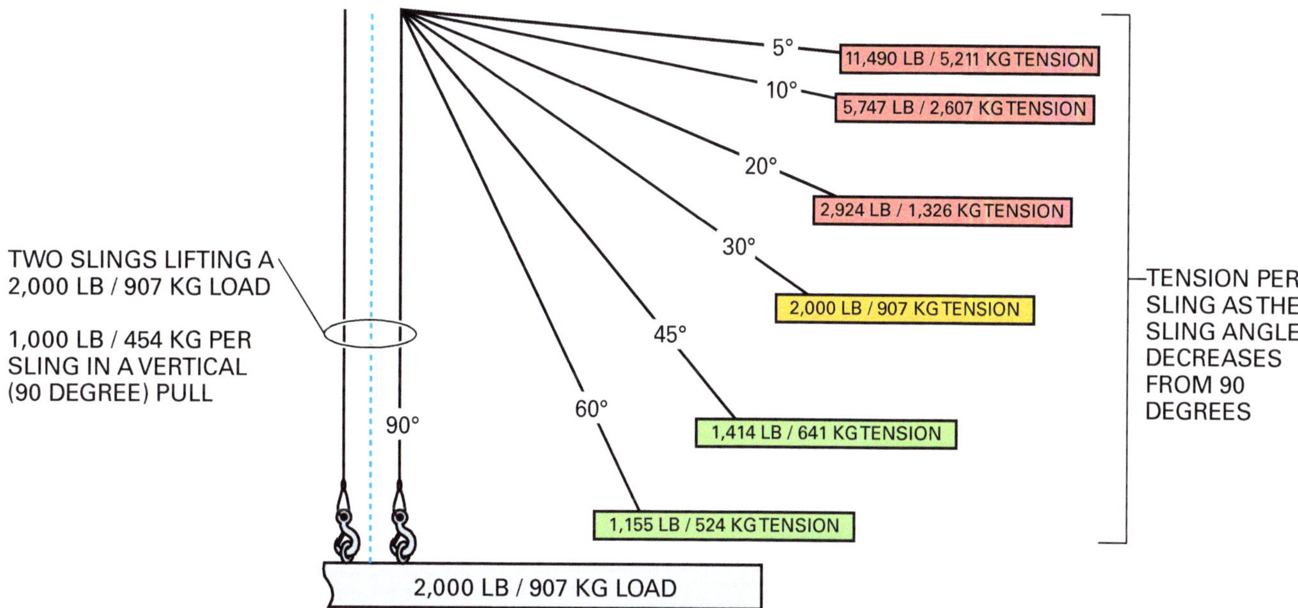

Figure 2 Sling tension as it relates to sling angle.

and the Crosby Links and Rings (*Figure 6*). Since standards and component specifications change from time to time, be sure to acquire and work with the latest edition of this aid.

There are two separate charts provided for the capacity of wire rope slings. One is for Extra Improved Plow Steel (EIP), and the other is for Extra Extra Improved Plow Steel (EEIP). The capacities shown, in tons, are based on the wire rope diameter and the type of hitch. Note that the two-legged sling and two-legged choker hitches show the load capacity at specific sling angles.

The Crosby Links and Rings chart shows the WLL for several different metals, based on the size of the link. Entries shown as N/A simply indicate unavailable sizes for that link. Note that this chart is to be used with slings at the common design factor of 5.

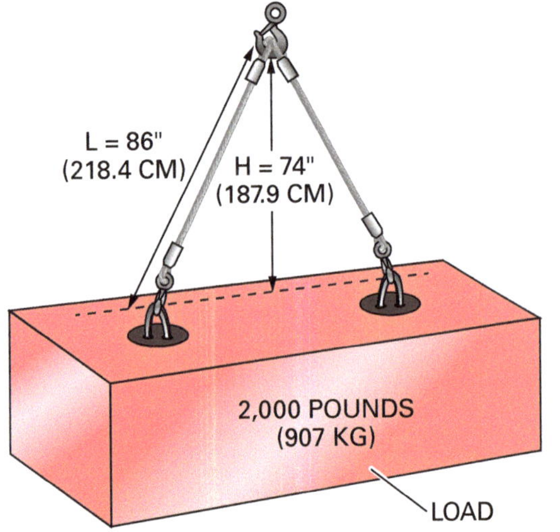

IMPERIAL VALUES	METRIC VALUES
Sling factor = L ÷ H	Sling factor = L ÷ H
Sling factor = 86" ÷ 74"	Sling factor = 218.4 ÷ 187.9 CM
Sling factor = 1.162	Sling factor = 1.162

Figure 3 Calculating the sling-angle factor (SAF).

1.1.2 Bridle Hitches

A bridle hitch consists of two or more slings that support the load attached to a common lifting point. The free ends of the bridles attach directly to the load at lifting lugs or eyes using some form of hardware. *Figure 7* shows a dual-sling bridle hitch.

The method used to determine and use sling-angle factors to reduce the capacity of a sling was covered earlier. That process results in finding the actual load (weight + tension) applied to a sling

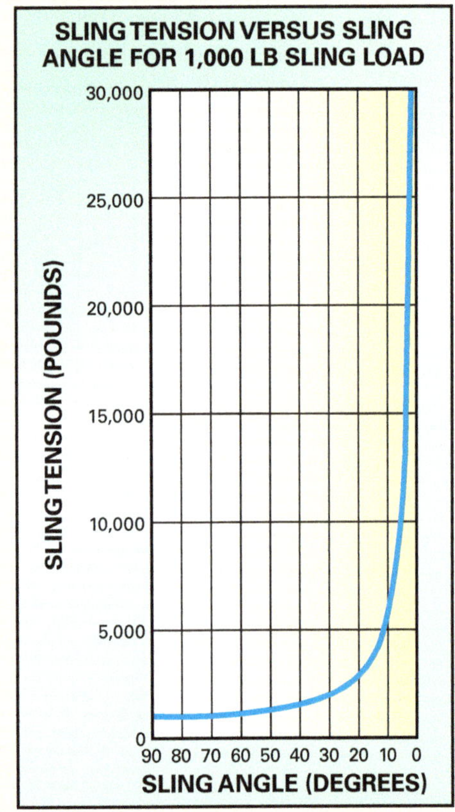

Tension Versus Sling Angle

Tension in a sling attached to a load is a geometric function of the portion of the load the sling is carrying and its sling angle. The graph shown here illustrates this relationship for a 1,000-pound sling load. Notice that as the sling angle approaches zero degrees, the tension required to support the sling load rises to extremely large values—much larger than the typical minimum breaking strength of sling materials.

at various angles. A somewhat more complicated method can be used to derate the sling's tagged or cataloged capacity. Both methods are used by individuals and organizations throughout the craft.

If you determine that a dual-sling bridle hitch evenly supports the load, calculate the available WLL for the hitch using the procedure that follows. If the slings are not evenly supporting the load, the formula used here will not provide an accurate result. First, determine the WLL for the sling's single leg vertical hitch from the manufacturer's information, or from the tag or label affixed to the bridle itself. This value is assigned the symbol W in the following formulas. Then, measure the length of a sling leg (L) between bearing points and the vertical distance between the sling's attachments (H), as previously shown in *Figure 3*. After obtaining these dimensions, use the following formula:

$$WLL = \frac{2W}{L \div H} \quad or \quad WLL = \frac{2W}{SAF}$$

Where:

WLL = dual-leg bridle's working load limit
W = vertical hitch WLL for one leg
SAF = sling angle factor for one leg of the bridle

> **CAUTION**
>
> These calculations do not include factors related to sling hardware that are considered by manufacturers in their ratings. This formula and process is not intended to guide the creation or fabrication of a complete sling assembly in the field.

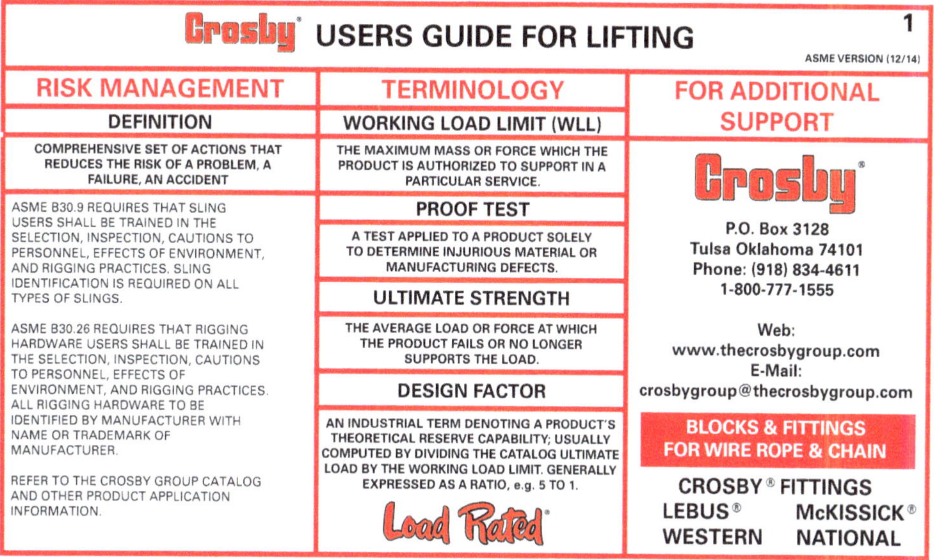

Figure 4 The *Crosby® Users Guide for Lifting*.

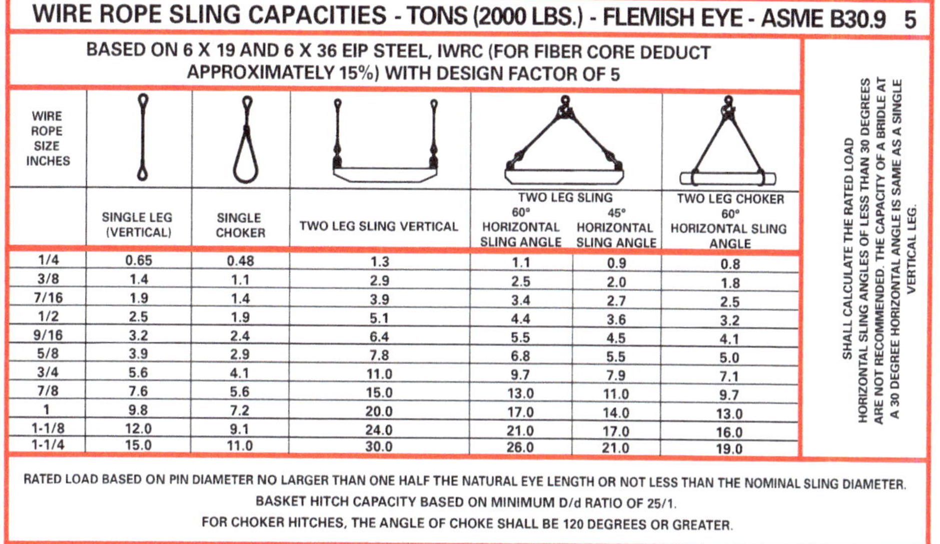

WIRE ROPE SLING CAPACITIES - TONS (2000 LBS.) - FLEMISH EYE - ASME B30.9   5

BASED ON 6 X 19 AND 6 X 36 EIP STEEL, IWRC (FOR FIBER CORE DEDUCT APPROXIMATELY 15%) WITH DESIGN FACTOR OF 5

WIRE ROPE SIZE INCHES	SINGLE LEG (VERTICAL)	SINGLE CHOKER	TWO LEG SLING VERTICAL	TWO LEG SLING 60° HORIZONTAL SLING ANGLE	TWO LEG SLING 45° HORIZONTAL SLING ANGLE	TWO LEG CHOKER 60° HORIZONTAL SLING ANGLE
1/4	0.65	0.48	1.3	1.1	0.9	0.8
3/8	1.4	1.1	2.9	2.5	2.0	1.8
7/16	1.9	1.4	3.9	3.4	2.7	2.5
1/2	2.5	1.9	5.1	4.4	3.6	3.2
9/16	3.2	2.4	6.4	5.5	4.5	4.1
5/8	3.9	2.9	7.8	6.8	5.5	5.0
3/4	5.6	4.1	11.0	9.7	7.9	7.1
7/8	7.6	5.6	15.0	13.0	11.0	9.7
1	9.8	7.2	20.0	17.0	14.0	13.0
1-1/8	12.0	9.1	24.0	21.0	17.0	16.0
1-1/4	15.0	11.0	30.0	26.0	21.0	19.0

RATED LOAD BASED ON PIN DIAMETER NO LARGER THAN ONE HALF THE NATURAL EYE LENGTH OR NOT LESS THAN THE NOMINAL SLING DIAMETER.
BASKET HITCH CAPACITY BASED ON MINIMUM D/d RATIO OF 25/1.
FOR CHOKER HITCHES, THE ANGLE OF CHOKE SHALL BE 120 DEGREES OR GREATER.

Figure 5 Wire Rope Sling Capacities chart (EIP slings).

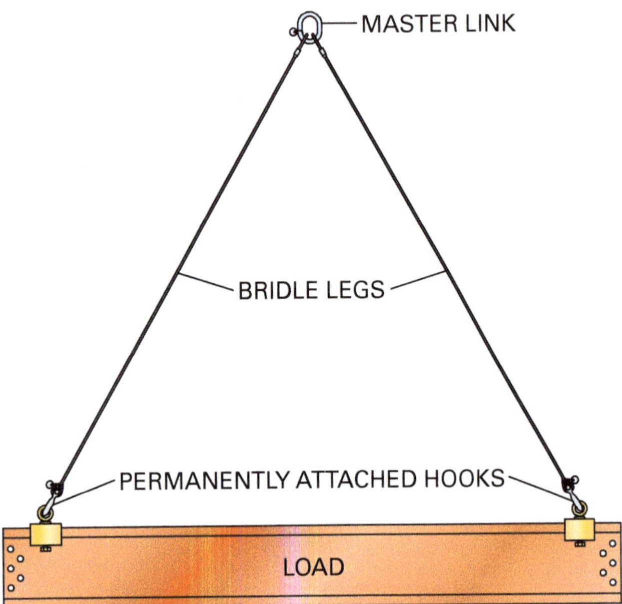

SIZE OF LINK IN INCHES	G-341 CARBON	A-341 ALLOY	A-342 ALLOY
1/2	2900	7000	7400
5/8	4200	9000	9000
3/4	6000	12300	12300
7/8	8300	15000	15200
1	10800	24360	26000
1-1/8	N/A	30600	N/A
1-1/4	16750	36000	39100
1-3/8	20500	43000	N/A
1-1/2	N/A	54300	61100
1-5/8	N/A	62600	N/A
1-3/4	N/A	84900	84900
2	N/A	102600	102600

SIZE	WORKING LOAD LIMIT JAW AND EYE 5/1 DESIGN FACTOR	WORKING LOAD LIMIT HOOK END FITTING. 5/1 DESIGN FACTOR
1/4	500	400
5/16	800	700
3/8	1200	1000
1/2	2200	1500
5/8	3500	2250
3/4	5200	3000
7/8	7200	4000
1	10000	5000
1-1/4	15200	N/A
1-1/2	21400	N/A

Figure 6 Crosby Links and Rings chart.

Figure 7 Dual-sling bridle hitch arrangement.

The formula reduces the vertical hitch capacities of each bridle leg to account for the sling angles. The rigger must then compare the calculated WLL to the load's weight to verify that the bridle will safely support the load. If the bridle has three legs, replace the factor 2 in the formula with 3; for a four-leg bridle, the factor is 4.

> **CAUTION**
>
> This formula is valid only for symmetrical loads. If there is any possibility that the sling legs are not supporting the load symmetrically, you must calculate the WLL for each leg of the bridle.

Rigging an asymmetrical load requires a more complex approach to calculate sling stresses. For example, *Figure 8* shows a custom, unequal leg bridle supporting a 24,000-pound angular load. The hook is directly above the load's center of gravity, so the load is balanced. The important dimensions, as shown on *Figure 8*, are defined as follows:

- L_1 – Length of sling A = 24.0'
- L_2 – Length of sling B = 9.3'
- D_1 – Horizontal distance from the sling A attachment point to directly below its master link bearing point = 13.5'
- D_2 – Horizontal distance from the sling B attachment point to directly below its master link bearing point = 5.7'
- H_1 – Height from the sling A attachment point to its master link bearing point = 19.9'
- H_2 – Height from the sling B attachment point to its master link bearing point = 7.5'

> **CAUTION**
>
> If the asymmetrical load's CG is not directly below the hook when making the calculations, the load will rotate when lifted. This can change the data used to make the sling tension calculations, making the results inaccurate. Irregular loads can create unexpected stresses. Always use a minimum 6:1 safety factor for sling/hardware selection.

The example shown calculates the tensions in the slings by taking into account the locations of the attachment points and the master link in relation to the load's CG. Note that the results have been rounded up to the nearest whole number.

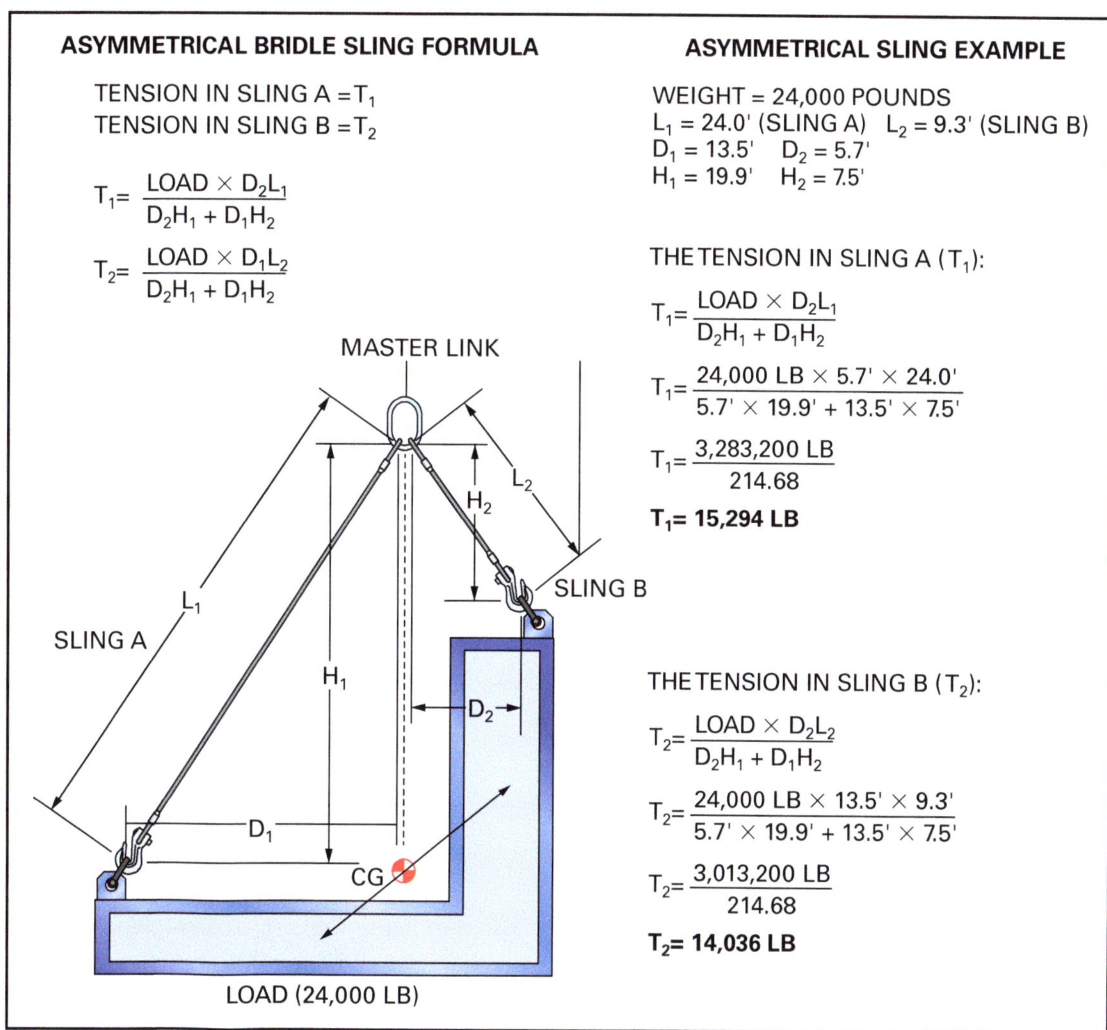

Figure 8 Asymmetrical bridle-sling example.

Generally, the closer horizontally an attachment point is to the CG, the greater the share of the load a sling carries for a given sling angle. The smaller the sling angle, the greater tension is in the sling. However, these two factors can interact in complex ways depending on the geometry of the hitch.

In practice, riggers must compare the resulting sling tension values against the vertical hitch WLL for each sling to verify that the slings have the capacity to lift the load.

1.1.3 Basket Hitches

A basket hitch cradles the load in the sling, and both ends attach to the hook. Basket hitches include single, double, and double-wrapped arrangements. A simple basket hitch cradles the load in the sling. A double basket hitch uses two basket hitches, either with a spreader bar or leading to a single point of attachment. A double-wrapped basket hitch wraps the sling around the load once before attaching to the hoist. Each type of basket hitch has its advantages and disadvantages, depending on the nature of the load. Use the following guidelines to determine the WLL of basket hitches:

- *Basket hitch* – For vertical legs, take the WLL of the single vertical hitch and multiply it by 2. For inclined legs, the formula is the same as that for the bridle hitch.
- *Double basket hitch* – For vertical legs, multiply the WLL of the vertical hitch (W) by 4. This hitch doubles the capacity of one basket hitch, and increases load stability. It may not be as secure on the load, however; if slack is allowed in the slings, the load becomes loose and can slide out of the hitch. Constant tension on the slings must be maintained. For angled legs, the basket hitch WLL formula is:

$$WLL = W \div SAF \times 4$$

> **NOTE**
>
> When using this formula, all four legs of the slings must carry the load equally.

- *Double-wrapped basket hitch* – The WLLs for double-wrapped basket hitches are the same as those for the double basket hitches. This hitch enhances load control.

Basket hitches differ from bridle hitches in that the sling material is in direct contact with the load. To avoid excessive local stresses on the sling material, *ASME B30.9* specifies a minimum ratio of load diameter to sling diameter (the D/d ratio). *Table 1* provides the *ASME B30.9* listed D/d ratios for various sling types. When these minimum values are maintained, the sling retains 100 percent of its WLL. If the sling must be used at a lower D/d ratio than the minimum, the sling must be derated according to the manufacturer's guidance. For example, if a 1" synthetic rope sling is wrapped around an object that is only 7" in diameter (a 7:1 ratio), then the manufacturer will derate it to a lower percent of capacity.

<table>
<tr><td>**CAUTION**</td><td>Each sling material has limiting D/d ratios specified in *ASME B30.9*. Riggers must derate basket hitch WLLs when D/d ratios fall below these limits.</td></tr>
</table>

1.1.4 Choker Hitches

A choker hitch is wrapped around the load by running the standing part of the sling through an eye, hook, shackle, or other ring-like fitting at the load end. The weight of the load tends to cinch the choker hitch around the load. Choker hitch types include single leg, double leg, and double-wrapped arrangements. The two parts of the sling forming the choke naturally tighten to approximately a 45-degree sling angle inside the choke (*Figure 9*), unless the load is unusually large or small.

When a choker hitch is used, riggers must derate the sling's vertical hitch WLL according to the manufacturer's guidance. If a given sling or its manufacturer doesn't have choker hitch capacity documentation, *ASME B30.9* provides general derating guidance based on the material of the sling, as shown in *Table 2*.

Under some circumstances, choker hitches may tend to tighten up under load. This can occur when lifting a bundle of loose round rod, for example, or when attempting to turn a load in the hitch. Cinching a choker hitch can create very small sling angles at the choke, which greatly increases the tension in the sling. For this reason, rigging professionals use the choke angle

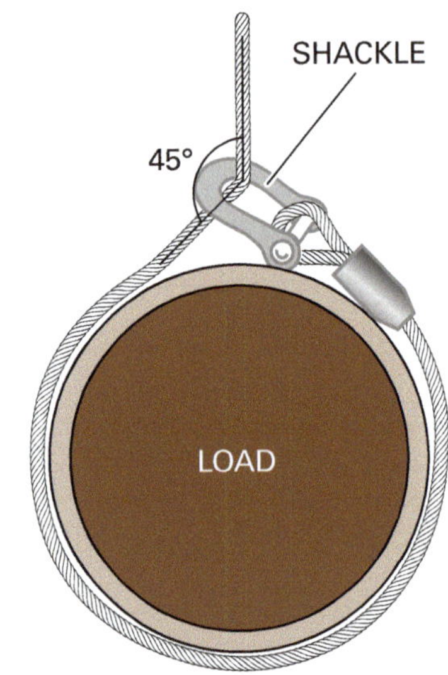

Figure 9 Choker hitch arrangement.

Table 1 Minimum Sling D/d Ratios

Sling Type	Minimum D/d Ratio
Alloy-metal chain	6
Wire rope (hand-tucked spliced eyes)	15
Wire rope (mechanical splice socket-type eyes)	25
Metal mesh	(none listed)
Synthetic rope	8
Synthetic mesh	(none listed)

Source: *ASME Standard B30.9-2014, Slings*

NCCER – *Intermediate Rigger*

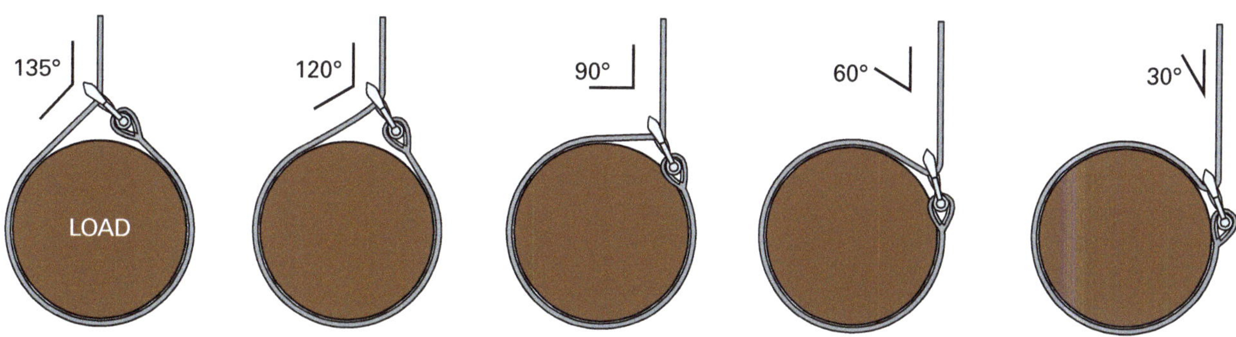

CHOKE ANGLE	CAPACITY %
OVER 120°	100
90° – 120°	87
60° – 89°	74
30 – 59°	62
0 – 29°	49

Figure 10 Choker hitch derating when cinched.

to determine the WLL of a sling in this condition. Measure the choke angle in the hoisted end of the sling where it bends, as shown in *Figure 10*.

> **NOTE**
>
> If using a sling that has no choker hitch capacity tag, first derate the sling's capacity according to *Table 2*. Then apply additional choke angle derating if the choke angle is less than 120 degrees, as shown in *Figure 10*.

Derating the sling's WLL from its vertical hitch rating in a choker hitch must take into account both the choke angle at the load and the sling angle of the hoisted end, when applicable. Use the following guidelines to determine the

Table 2 ASME Choker Hitch Derating Without Documentation

Sling Material	Maximum WLL
Wire rope (cable-laid)	70%
Synthetic rope	75%
Wire rope (braided)	75%
Alloy metal chain	80%
Polyester roundsling	80%
Synthetic webbing	80%
Metal mesh	100%

Source: *ASME Standard B30.9-2014, Slings*

WLL of choker hitches (the abbreviation W in the formulas is the single vertical hitch WLL):

- *Single leg choker hitch* – Measure the choke angle and apply the derated capacity to the sling's vertical hitch WLL:

  WLL = W × choke angle rated capacity

- *Double leg choker hitch* – When using more than one sling with a choker hitch, use the single leg choker hitch formula combined with the sling angle calculations for bridle hitches. The following formula applies:

  WLL = W × (choke angle rated capacity) ÷ (leg SAF) × 2

- *Double-wrapped choker hitch* – The values for the double-wrapped choker hitch are the same as those for the single choker hitch or the double choker hitch, depending on the configuration.

It is easy to overload a choker hitch if you do not consider the many capacity-reducing factors for every application.

1.2.0 Special Equipment for Moving Loads

Riggers must use special equipment to move loads horizontally, as well as for short distances vertically. This is often required inside buildings where a crane cannot assist. Equipment commonly used for this purpose includes jacks, tuggers, grip hoists, skids, and rollers.

1.2.1 Jacks

Riggers use jacks to raise or lower a load over relatively short distances. Hydraulic jacks (*Figure 11*), which multiply the applied force by utilizing the pressure of an enclosed liquid, are the most common type used for heavy lifting. They come in a variety of configurations, including bottle jacks, hand- and electric-pump jacks, hydraulic spreaders (*Figure 12*), and pullers. Inflatable jacks are also used.

Other types of jacks available for rigging tasks are single-acting, low-profile cylinders with screw and locking features (*Figure 13*), as well as double-acting, center-hole cylinders for push-pull applications. Manufacturers combine hydraulic cylinders with pumps, hydraulic hoses, and other accessories to form jacking systems (*Figure 14*).

To select an appropriate jack, riggers must consider the weight, distance, and points of contact with the load. These factors determine the type of pumps, cylinders, or other jacking systems required. A good rule of thumb is to select devices with a capacity of about 50 percent (or 1.5 times) greater than the load. This will prevent overloading and damage to the jack, and will also serve as reserve capacity in case the load is underestimated.

When performing jacking operations, fully support the cylinders on a firm, non-sliding foundation. Use a sacrificial material, such as scrap wood, between the tip of the ram and the load. Always use hardwood for blocking, and follow the load up or down with the blocking. Plan to have the thickness of the blocking match the extension of the ram. Make sure there is adequate space for jacks and blocking, and jack and block at locations that can withstand the forces exerted. This refers to the supporting foundation as well as the points on the load itself.

> Prevent metal-to-metal contact by using wood or similar material between the jack and a metal load.

While jacking or blocking, stand back occasionally and check the overall operation for stability.

Figure 12 The "jaws of life" is an example of a spreader jack.

Figure 11 Hydraulic jacks.

Figure 13 Hydraulic cylinders.

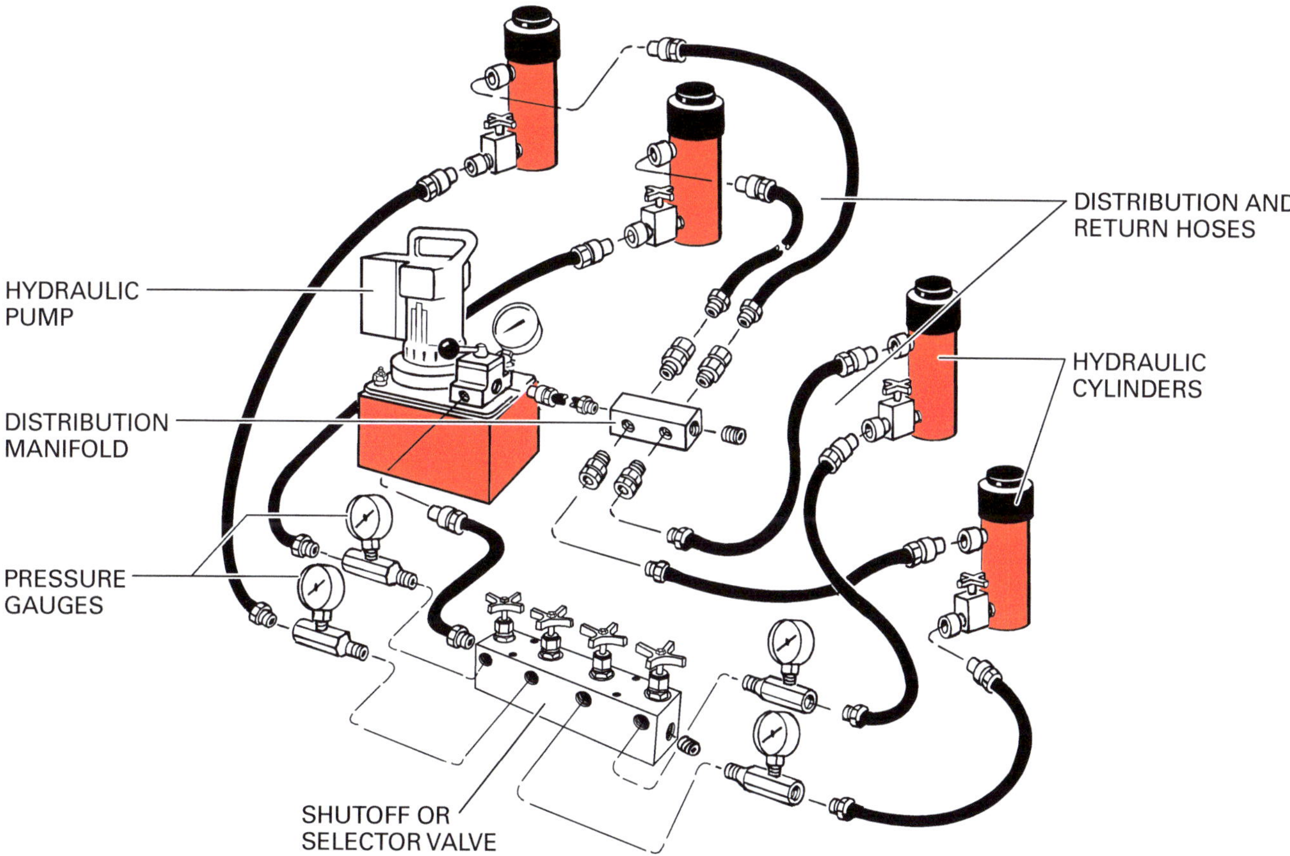

Figure 14 Hydraulic jacking system.

If it doesn't look stable, it probably isn't. Take steps to stabilize the load to prevent an accident, even if it means starting over with a different approach.

1.2.2 Grip Hoists

The grip hoist (*Figure 15*) is a wire rope hoist used for lifting and pulling. In practice, manufacturers do not rate grip hoists for vertical lifts, and riggers should not use them that way. The term *hoist* can be misleading—these devices are meant to be used for horizontal pulling only. However, they can be used with tackle systems, where pulleys are used to change the direction of the pull, allowing lifts to be done with them.

Various models of grip hoists accept wire rope from $\frac{5}{16}$" to $\frac{5}{8}$" (8–16 mm) in diameter. The grip hoist is not limited to a certain length of pull, because it continuously accepts wire rope from an outside source. The mechanism contains gripping jaws to grasp the rope when you pull the hand lever. The jaws automatically hold this position when not actively pulling the rope.

The grip hoist has one attachment point mounted on the end of the housing to secure the device. Before operation, riggers must secure the grip hoist to an anchorage point that is strong and stable enough to support the intended load. The rigger feeds the free end of a wire rope attached to the load into the hoist and out the back.

> **CAUTION**
>
> Always check the condition of the rope prior to any lifting or pulling.

> **WARNING!**
>
> Failure to secure the hoist to an anchorage point strong enough to support the load can cause serious damage to the structure and/or load, possibly resulting in injury or death.

1.2.3 Skids

Skids are long timbers placed lengthwise under heavy loads in the direction of the intended movement. They perform the following functions:

- Distribute the load's weight over a greater area, reducing ground pressure
- Provide a relatively smooth, firm surface over which equipment can be moved
- Provide a runway surface for rollers, when used

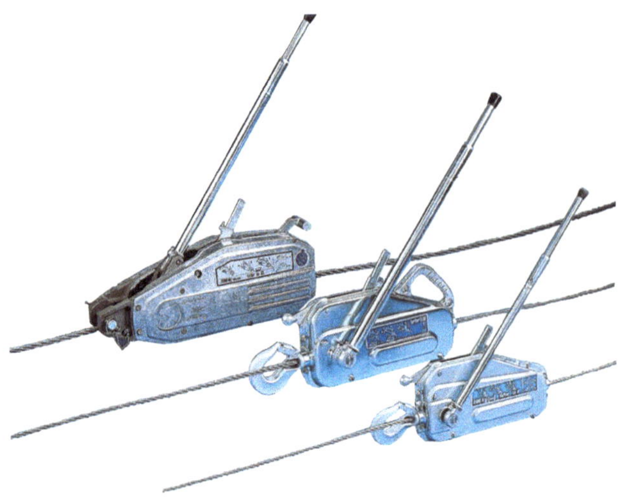

Figure 15 Grip hoists.

For most operations, oak timbers 2 or 3 inches (5 or 8 cm) thick and about 15 to 20 feet (5–6 m) long make effective skids. The length may need to be adjusted to accommodate the space available and the distance of the move. When using skids, it is good to stagger the joints between lengths so that no two sets of joints are at the same point.

Skids can be made from a variety of materials. In some cases, the floor itself is used as the skidding surface. Friction will result between the materials as they move, related to the type of materials and the weight of the load. Engineers refer to the friction factor related to the movement of materials in contact with each other as the coefficient of friction (CF). The coefficient of friction falls between zero and the whole number 1 for non-adhesive materials. Coefficients for common materials are shown in *Table 3*.

For example, assume you are moving a piece of steel machinery into position, skidding across level wooden skids. As shown in *Table 3*, the CF is 0.30. If the machinery weighs 2,800 pounds, determine the line pull as follows:

$$\text{Line pull} = 2{,}800 \text{ lb} \times 0.30$$
$$\text{Line pull} = \textbf{840 lb}$$

If the load is simply pulled across a concrete floor without using skids, the CF rises to 0.60. The required line pull is then much higher—1,680 pounds. If it is practical to do so (primarily for very heavy loads, due to the labor and expense), PTFE (Teflon™) can be applied to the skids and the load, reducing the CF to 0.04. This results in a line pull of a mere 112 pounds.

Irregularities in the load's bearing surface, such as sharp edges, can increase the line pull significantly. Because of the many variables in surface irregularities, it is difficult, if not impossible, to accurately account for these irregularities

Materials		Coefficient of Friction (Static)
Concrete	Concrete	0.65
Concrete	Metal	0.60
Concrete	Wood	0.45
Concrete	Rubber	0.90–1.0
Steel	Steel	0.74
Steel	Steel (lubricated)	0.15
Steel	Aluminum	0.60
Steel	Cast iron	0.25
Wood	Wood (species-dependent)	0.25–0.50
Wood	Metal	0.30
Wood	Manila rope	0.40
Teflon™ (PTFE)	Teflon™ (PTFE)	0.04

mathematically. Almost any form of lubrication applied to the surfaces will significantly reduce the line pull. Even a coat of glossy paint on a steel surface will reduce the line pull to some degree. Therefore, calculating the line pull using the CF is merely a reasonable estimate of the actual pull required.

1.2.4 Rollers

The practice of using rollers to move loads is as old as the rigging craft itself. Riggers can use pipe or round-bar rollers over skids or a smooth surface for some loads (*Figure 16*). However, rollers are often used directly on the surface beneath, without skids, when the surface is smooth enough. The surface must also be clean—even a small pebble can stop a rolling pipe instantly. In general, moving equipment on rollers requires considerably less effort than skidding and other means.

The rollers must be long enough to pass under the load completely, from one side to the other. As the load moves along, workers bring the free rollers at the rear back to the front and place them under the approaching load. To make a turn, they tap the outer ends of the front rollers slightly in the direction of the turn, and adjust the rear roller angles slightly in the opposite direction. The line pull required for round rollers on a flat surface for any materials is usually less than 10 percent of the load weight. For rubber or metal rollers on a firm, smooth surface, the line pull may be as low as 1 percent.

Sturdy mechanical rollers have become more popular for moving large loads. Jacks or pry bars raise the load to provide clearance, then the roller assemblies are slid under the load. *Figure 17* shows examples of mechanical rollers. Sizes are available to move loads ranging up to thousands of tons. Mechanical rollers generally require a rigid, flat, supporting surface, such as plate metal or smoothed concrete. The surface must be able to bear the weight of the load.

Safety is the most important consideration when moving large loads. Riggers and workers must observe the following basic precautions:

- Avoid pinch points.
- Chock loads properly when leaving them on rollers to prevent unintentional movement.
- Use restraining lines and anchor points rated for the expected load when moving loads on a downward slope.
- Be sure to secure the load to the skid or dolly when transporting it.
- Be certain the dollies and rollers have sufficient capacity to carry the load.
- Make sure the surface over which the load will move is as smooth and level as possible.

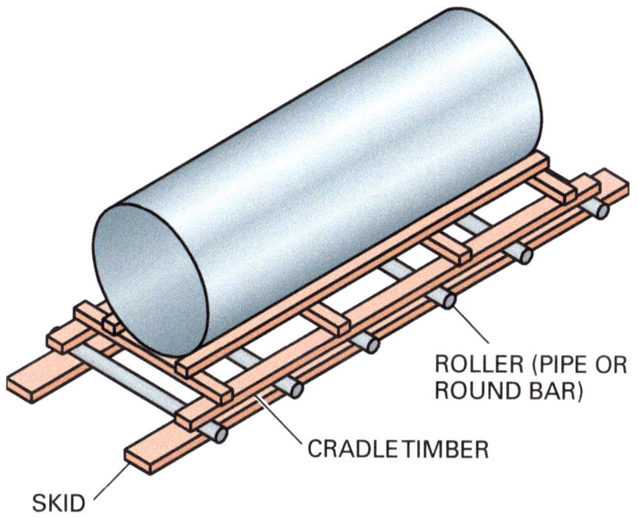

Figure 16 Moving a load on pipe and skids.

(A) MECHANICAL ROLLERS

(B) ROLLERS IN USE

Figure 17 Mechanical rollers.

- Determine the center of gravity to ensure stability before moving a load on rollers or dollies.
- Make sure there is an adequate number of rollers under the load to safely support it and maintain stability.
- When skidding over dirt or unstable surfaces, use timbers, steel mats, or steel plates.
- Apply moving force to the skid, not the load.

1.3.0 Block and Tackle

The block and tackle is a basic lifting system. A block is comprised of one or more sheaves (pulleys) fitted into a frame with a hook attached to the top. The tackle consists of the rope and any hardware connected to the block(s). Some have a brake that holds the load once the hoisting line, referred to as the *hauling line*, is released. Those without a brake require a continuous pull on the hauling line, or it must be tied off to a secure point.

There are two basic types of block and tackle rigs: simple and compound. A simple block and tackle consists of one sheave and a single line (*Figure 18*). It is used to lift or pull light loads. The load-line hook is attached to the load, and the load is lifted by pulling the hauling line. The block must be attached to a building structure or other support that provides adequate capacity to support the load, tackle, and other rigging components. Adequate strength and stability of the supporting structure is crucial and must be evaluated carefully.

A compound block and tackle (*Figure 19*) uses a second block. It has a fixed upper block that is attached to the building structure or other support, and a traveling lower block that is attached to the load. The more sheaves the two blocks have, the more parts of line the assembly can accommodate. More parts of line provide a mechanical lifting advantage, as discussed in the next section.

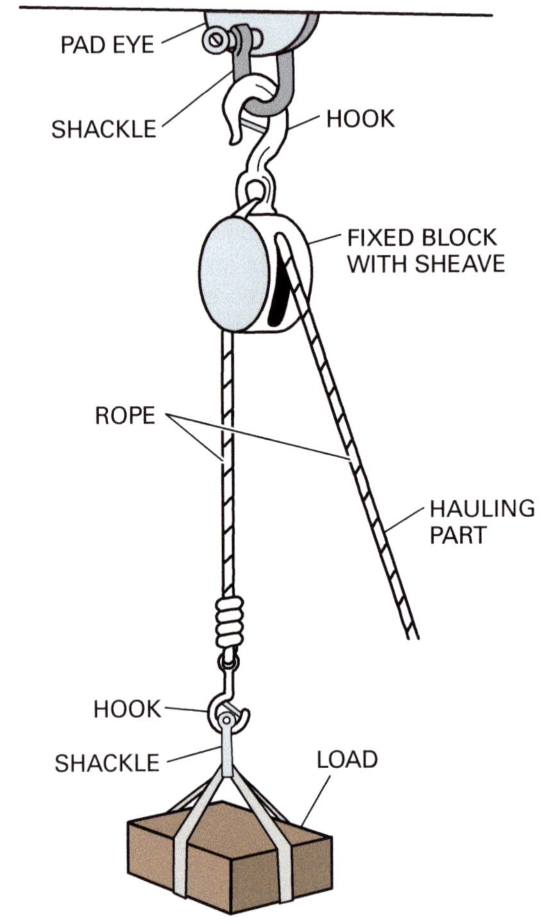

Figure 18 Simple block and tackle.

NCCER – *Intermediate Rigger*

Figure 19 Compound block and tackle.

1.3.1 Applied Loads and the Mechanical Advantage

Using a block and tackle for lifting a heavy object is often done to gain a mechanical advantage. The simple block and tackle in *Figure 18* provides no mechanical advantage. In other words, lifting a 100-lb load requires a 100-lb line pull. Simple block and tackle is generally used only to change the direction of the pull when a mechanical advantage isn't necessary for the task.

Compound block and tackle is used when the load is heavier and a mechanical advantage is needed. You aren't going to lift a 1,000-lb load without some sort of advantage. The number of parts of line reeved between the blocks determines the extent of the advantage. The following simple equation is used to determine the hoisting line pull needed to lift a load:

Line pull = load weight ÷ parts of line

> **NOTE**
>
> The weight of the rigging components must be accounted for in the load weight. For the sake of clarity, the weight of the rigging components is not considered here.

The hauling line is not a part that contributes to any mechanical advantage—only the parts of line between the two blocks are considered. *Figure 20* shows how a block reeved with two parts of line affects line pull. Using a block with four parts of line for a load of 300 lb would require a line pull of only 75 lb. This is clearly an advantage, and a line pull that can easily be generated by an average worker.

However, this advantage is offset by an increase in the amount of line required. With two parts of line, the required line pull for a load is halved, but the amount of line that must be hauled in doubles. In other words, for every foot the load is lifted, two feet of hauling line must be hauled. If there are five parts of line, the line pull is only $\frac{1}{5}$ of the load weight, but five feet of line must be hauled to lift the load one foot.

A block and its anchorage experiences a far greater load than you might expect. Refer to *Figure 21*. The block and anchorage point not only experiences the load weight, but also the line pull. For a

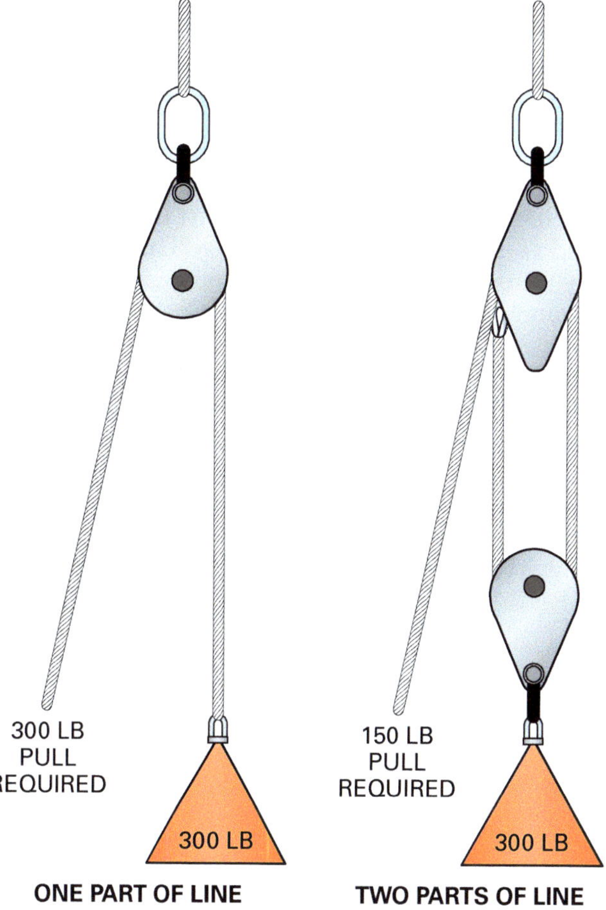

Figure 20 Line pull for one and two parts of line.

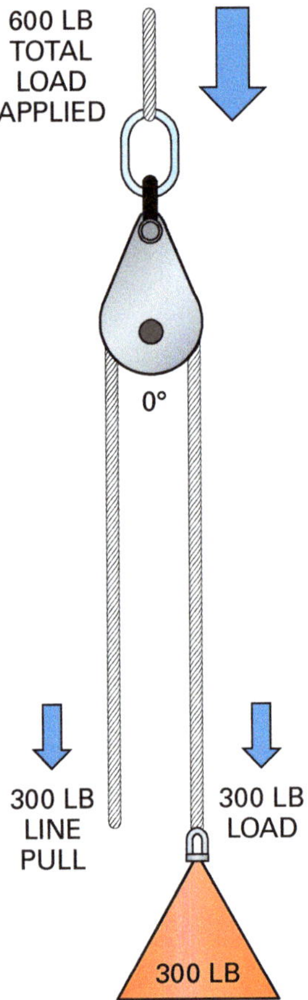

Figure 21 Load applied to the block when hoisting.

300-lb load hoisted on a single part of line, a 600-lb load is applied to the block as well as the anchorage. This clearly shows the importance of selecting an anchorage point that can support the applied load. Accidents often occur when those not properly trained attempt to lift loads with blocks or other types of hoists, anticipating that only the load weight is applied to the anchorage point. Compound blocks also experience additional stress from line pull, but to a lesser extent since the line pull per pound of load weight is reduced. However, the added line pull remains a substantial additional load on the block and anchorage.

In *Figure 21*, note that the two lines—the hauling line and the load line—are perfectly vertical. in practice, however, the hauling line is often at an angle to the block. The load line may also be at an angle. In this simple example, once the load leaves the ground, it will be vertical. However, when blocks and pulleys are used for multiple changes in direction, the load line can be at a variety of angles.

When one or both lines are at an angle other than vertical, the load applied to the block and anchorage point is reduced. The total load applied is determined by the following equation:

Total load applied = line pull × angle factor

The angle factor is a multiplier used to help determine the total load applied. It is based on the angle formed by the hauling line and the load line. In *Figure 22*, the angle formed by the two lines is 40 degrees. *Table 4* provides the angle factors for angles from 0 to 180 degrees. It shows that the angle factor for 40 degrees is 1.87. Therefore, the total load applied to the block and anchorage point for this example is determined as follows:

Total load applied = line pull × angle factor
Total load applied = 300 × 1.87
Total load applied = **561 lb**

Notice in *Table 4*, as you approach a 180-degree angle between the ropes, the total load applied drops rply. At 180 degrees, the rope is merely passing through the block in a straight horizontal line. Therefore, no real load is being applied to the block.

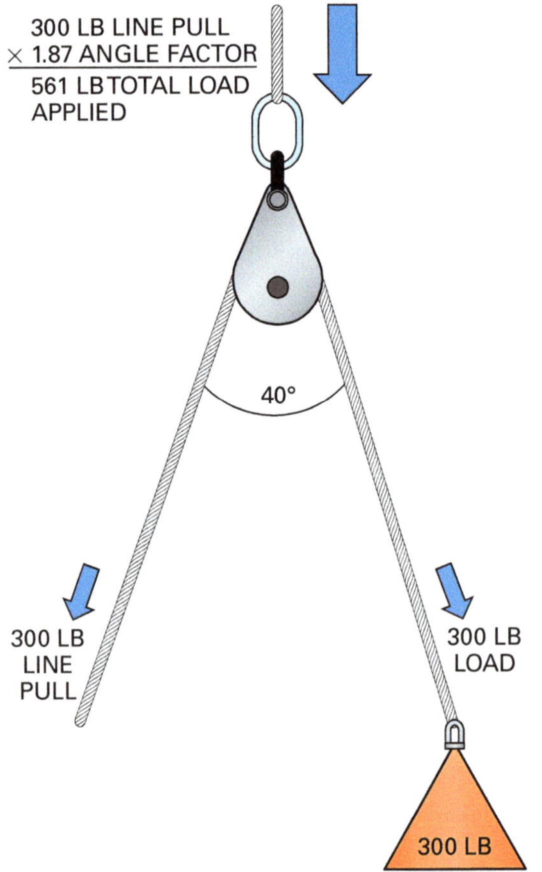

Figure 22 Applying an angle factor to determine the total load applied.

Table 4 Single Line Block Angle Factors

Angle	Factor
0°	2.00
10°	1.99
20°	1.97
30°	1.93
40°	1.87
50°	1.81
60°	1.73
70°	1.64
80°	1.53
90°	1.41
100°	1.29
110°	1.15
120°	1.00
130°	0.84
140°	0.68
150°	0.52
160°	0.35
170°	0.17
180°	0.00

In some cases, the direction of line pull must be changed several times. In *Figure 23*, the direction of pull is changed three times using blocks. The total load applied to each block can be calculated as shown in the diagram; the resulting loads have been rounded up to the nearest pound. The 150-lb line pull shown to the right is likely to actually be slightly higher due to the friction of sheave bearings and other small contributing factors. Friction factors are not generally considered, since they are usually insignificant. However, worn or defective sheave bearings can increase the effort required.

If the required line pull is more than riggers can safely sustain, an anchored winch or tugger may be required. All the blocks and hardware used must be selected carefully to ensure the applied loads are within their WLL.

1.3.2 Reeving Block and Tackle

Block and tackle uses the same basic reeving patterns as those used with crane load blocks. Refer to *Figure 24*. The hauling line should always be reeved across the center sheave, or on the sheave next to the center when an even number of sheaves exists. Reeving the hauling line on a sheave at or

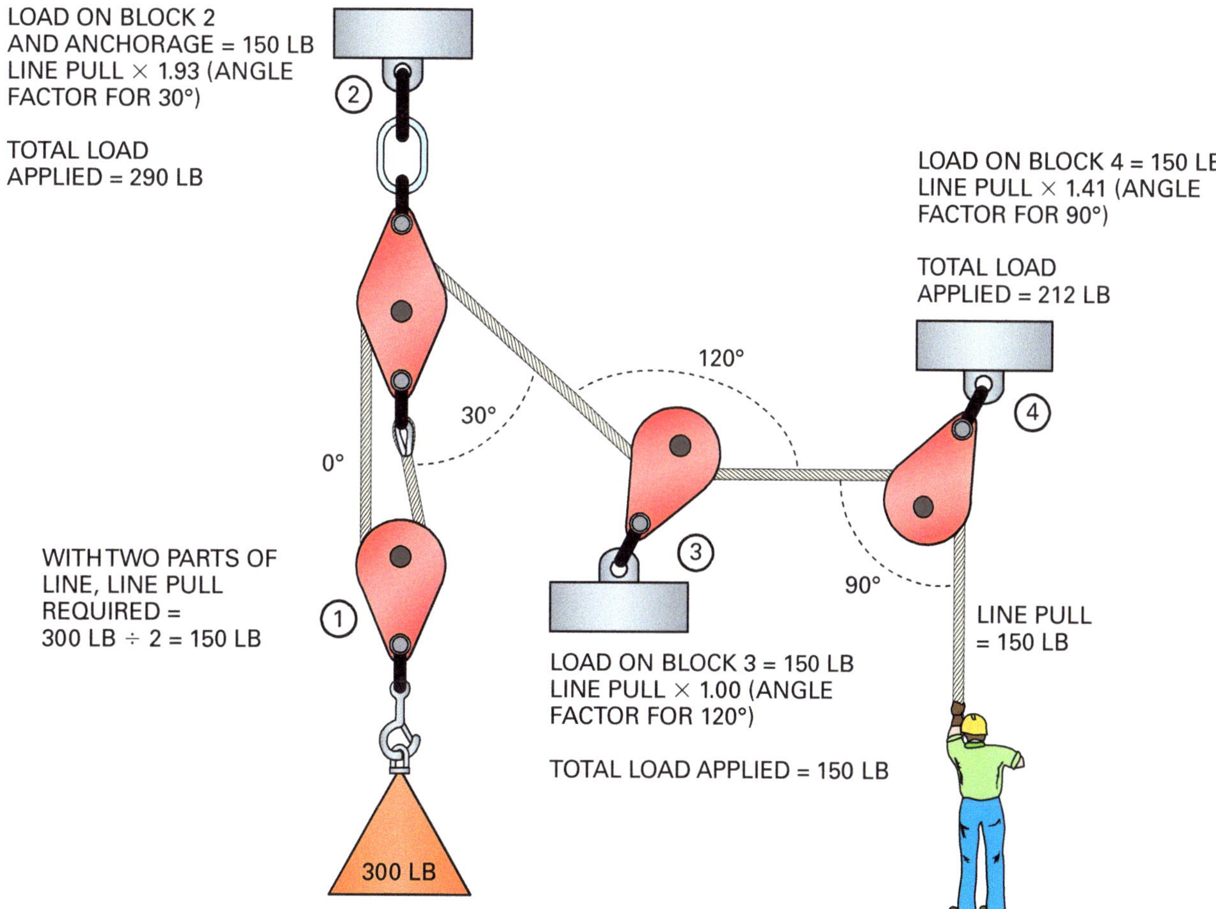

Figure 23 Applying angle factors to multiple changes in line pull direction.

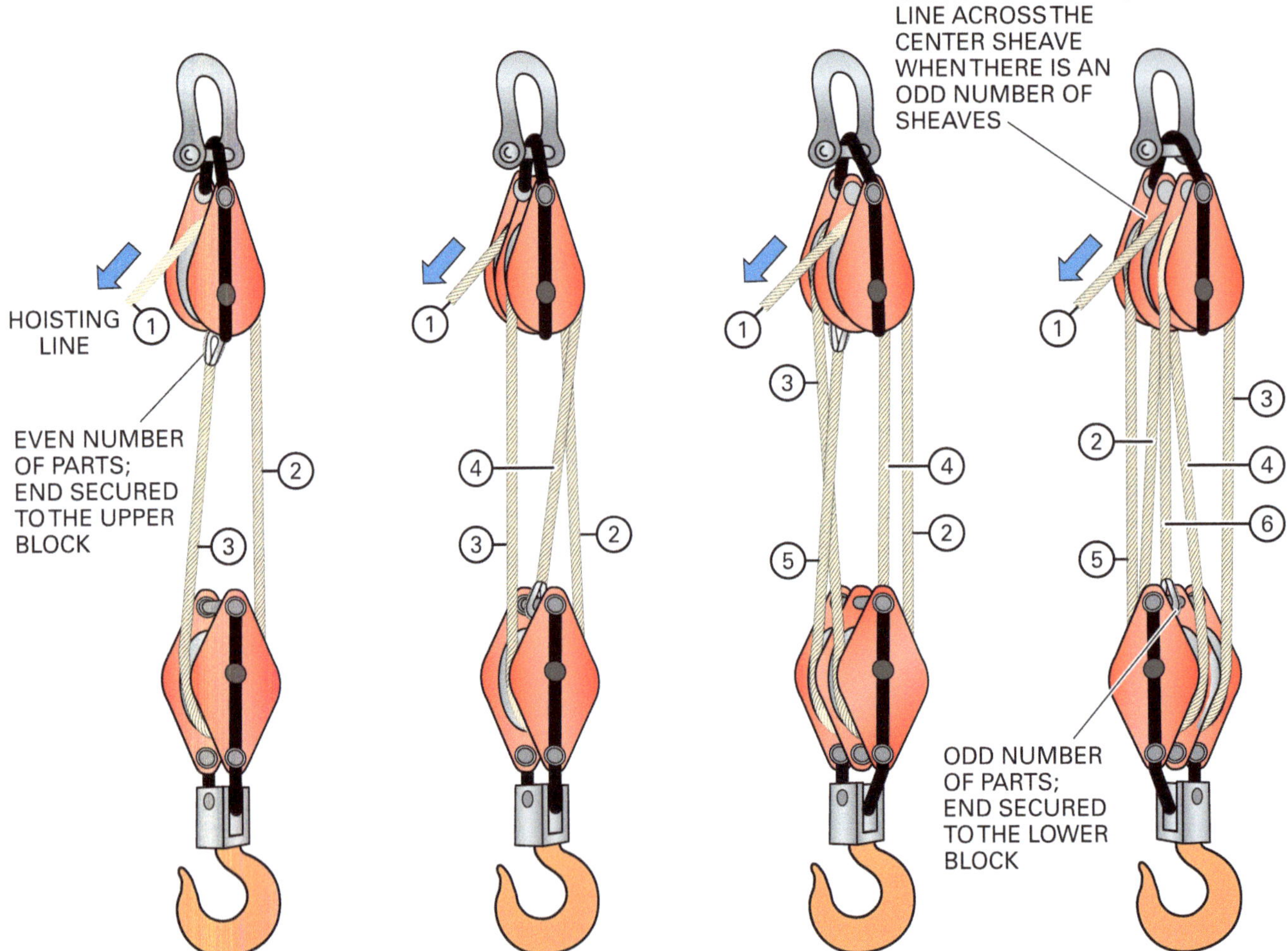

Figure 24 Block and tackle reeving.

too near one side of the block causes the block to turn to one side when line pull is applied. For cranes, this results in boom torque. When there are multiple parts of line, it is also important to reeve the blocks symmetrically.

1.4.0 Inverting Loads

Riggers must often invert loads onto their side, or fully invert them (flipping the top and the bottom). When outdoors, a crane can often assist. In areas that cannot be accessed by a crane, riggers must use block and tackle, ratchet hoists, and chain falls to invert a load.

When working with multiple hoists, it is best to have a single individual supervise the operation, with one or more workers operating individual hoists. Attempting to complete many inversion tasks alone is not smart or safe. Many organizations require that an appropriate amount of manpower be on hand, with a single individual directing and overseeing the operation from multiple vantage points.

Riggers may choose to use ratchet hoists, chain hoists, or block and tackle for inversion tasks. Come-alongs cannot be used for vertical lifting.

> **WARNING!**
>
> Never use a cable-equipped come-along or grip hoist for vertical lifting. Both are designed for horizontal pulls only. Although the cable itself may have plenty of load capacity, the mechanisms are not designed to effectively snub and safely hold a suspended load.

1.4.1 Inverting a Load with a Single Hoist

Some loads can be turned using a single hoist or hook. However, it is important to consider the load CG when using a single hook. If the hoist connection is below the CG, the load will become unstable as the lift proceeds, and the load will likely tip over and fall to the floor. If the hoist connection is in the same horizontal plane as the CG, the load may also tip and strike the floor, but once the load breaks contact with the floor, it will swing in balance. For turning a load on its side or inverting it, always lift from an anchorage point above the CG.

Refer to *Figure 25*. Note the location of the CG on the load, relative to the attachment point. To turn the load 90 degrees and lay it on its side, proceed as follows:

Step 1 Ensure that the anchorage point for the hoist, as well as the hoist itself, is sufficient for the calculated load (including the added stress from sling angles) and safe for use. Attach the hoist well above the CG, on the side opposite that on which you want it to lay (opposite of what will become the bottom).

Step 2 Lift the load slowly with the hoist. As it rises, it will naturally lean toward the side where the CG is located. When only a small portion of the load weight remains on the floor, allow it to continue tipping in the desired direction while lowering the hoist. Assist by hand to push the load over center if necessary.

Step 3 The load can usually be lowered to the floor with limited management once the hoist is again carrying a significant amount of weight. Unique, asymmetrical loads are more challenging to control.

Knowing the load's CG is critical. It is important to remember that the CG plays an important role in the selection of rigging connection points and the management of the load. When the load is near vertical, with only a small amount of load weight remaining on the floor, the load may turn or otherwise move unexpectedly. Maintaining a mental picture of the CG location throughout the operation will help you to predict what the load is likely to do.

1.4.2 *Inverting a Load with Two Hoists*

Some loads may require more than one hoist to invert. As the inversion progresses, there is a point where the one hoist that remains connected throughout the process must carry the entire load. Ensure that all rigging and hoists have the capacity for the loads imposed on them.

To invert a load as described here, you will need enough room above the operation to accommodate the hoists, slings and hardware, and the total height of the load once it is perpendicular and slightly off the floor.

The following steps describe how an uncomplicated inversion can be done (refer to *Figure 26*):

Step 1 Rig the load with the proper slings at acceptable sling angles. Calculate sling stresses before beginning. The points of load connection are best on the same horizontal plane as the load CG. This helps ensure the load is balanced and more easily managed once it becomes vertical.

Step 2 Lift the load slowly and evenly, using both hoists at the same pace. Raise the load until it is off the floor.

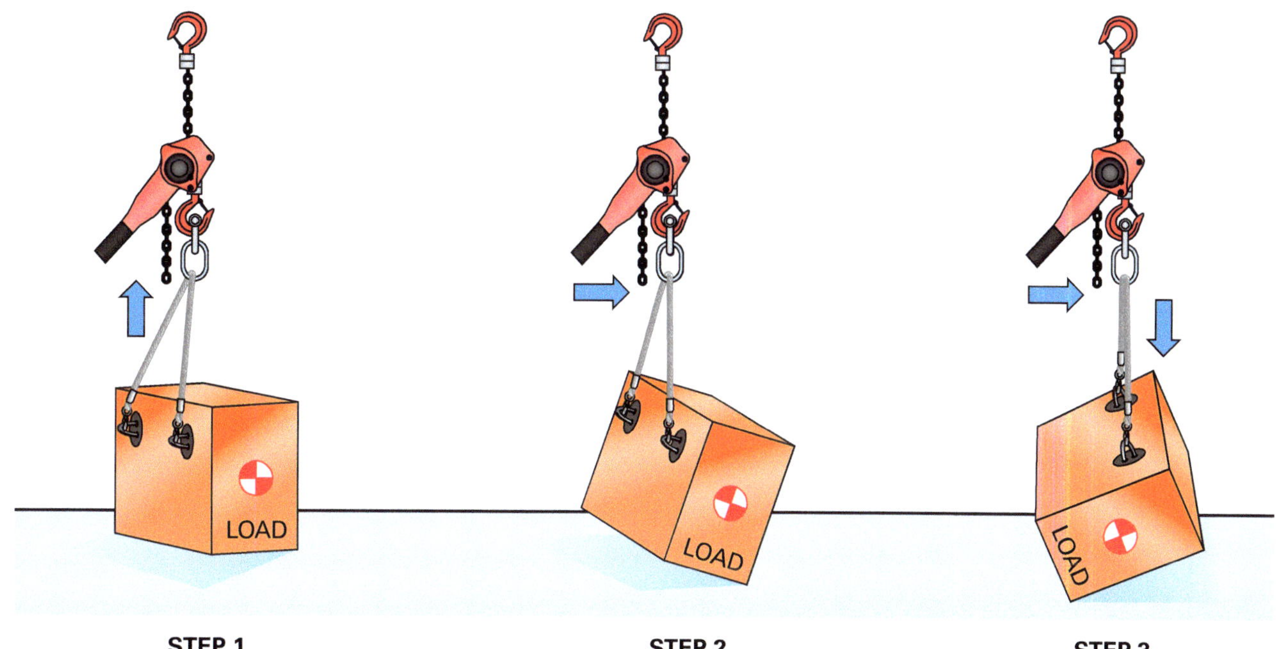

Figure 25 Inverting a load 90 degrees.

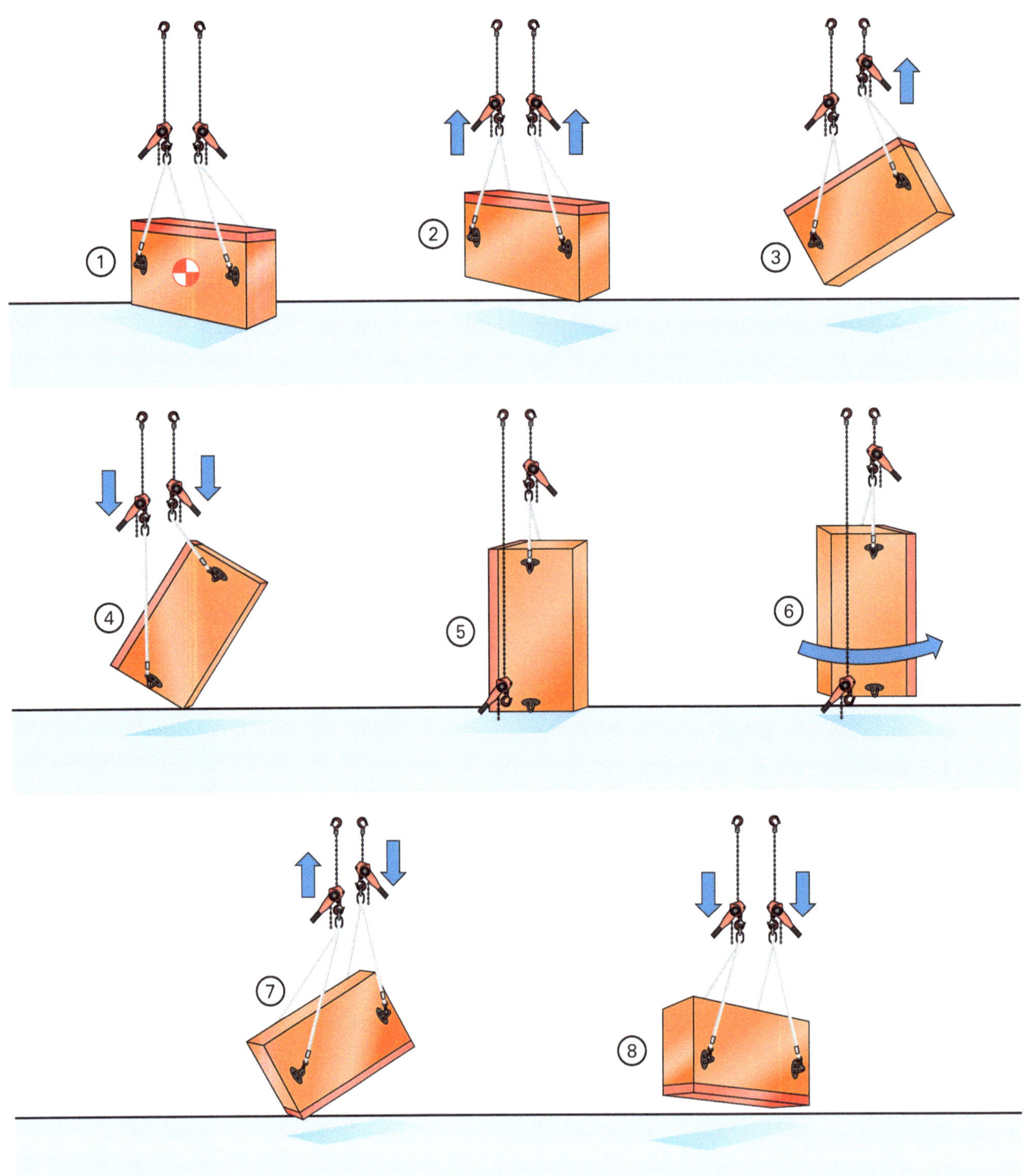

Figure 26 Inverting a load with two hoists.

Inverting a Load with a Choker Hitch

Some loads can be inverted using a choker hitch. Be sure to choke the load as shown here. Otherwise, when the load is inverted, the hitch will be difficult to release.

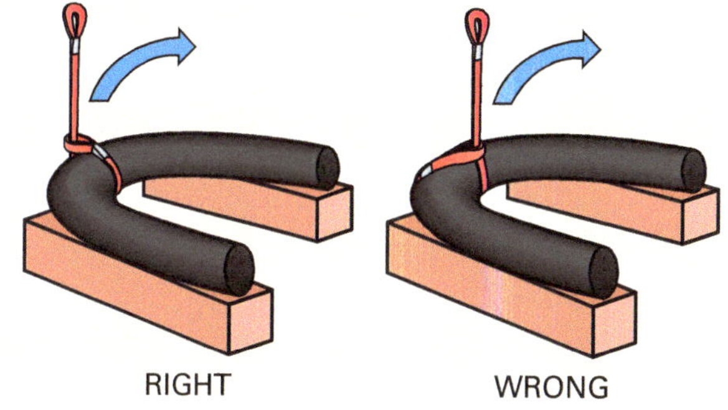

Step 3   Now using one hoist, continue raising one end of the load. If the bottom of the load touches the floor as it rotates, raise the other hoist slightly to keep it off the floor.

Step 4   Once the load is near vertical, lower both hoists as needed for it to touch the floor.

Step 5   Use one hoist to help raise the load until it stands up. The hoist to the lower connections can now be disconnected. The other hoist remains connected throughout the process.

Step 6   Raise the load just enough to rotate it 180 degrees while suspended from the hoist. Then lower it back to the floor.

Step 7   Reconnect the second hoist to the lower connection points. Raise the hoist slightly, placing at least some load on it to keep the slings taut. The load can remain in contact with the floor, or be kept suspended from both hoists for the next step.

Step 8   Now lower the upper end of the load to the floor, or until the load is suspended level. Then both hoists can be used together to lower it the remaining distance to the floor.

These basic approaches to turning and inverting a load may require some field modification, for a variety of reasons. As you participate in rigging tasks like these, you will gain confidence and better understand how to adjust the process to fit the circumstances, with safety always the priority.

1.4.3 Drifting a Load With Two Hoists from a Single Anchorage

Another method of fully inverting a load uses a single anchorage point and two ratchet hoists. The single anchorage point is usually occupied by a chain hoist, and two ratchet hoists are then attached to the chain fall. For unique situations, additional ratchet hoists may be involved. As always, it is essential to ensure that the anchorage point can support the load weight as well the weight of all rigging and any added stress from angular pulls.

Step 1   From a suitable anchorage point above, suspend a chain hoist. Use master links or appropriate shackles to connect the two ratchet hoists to the chain. The hooks of the ratchet hoists should *not* be directly attached to the hook of the chain hoist. Refer to *Figure 27*. Slings and other hardware are then used to connect the ratchet hoists to the load. Use softening devices and sling guards to protect the load surface finish and the slings from sharp edges.

Figure 27   Preparing a U-shaped load for inversion.

Step 2 Use the chain hoist to raise the load off the floor. Once it is sufficiently clear of the floor, use one ratchet hoist to raise one end of the load until it is vertical or near-vertical. See *Figure 28*. Once in this position, this same ratchet hoist must suspend the entire load; the second ratchet hoist is loosened and then disconnected from the load altogether.

Step 3 The second ratchet hoist is then moved to the opposite side of the load and reconnected. While the load remains suspended vertically from the first ratchet hoist, the second ratchet hoist is operated to raise the low end and begin returning the load to the horizontal position. See *Figure 29*. The first hoist is simultaneously lowered as needed to bring the load back to horizontal, with the top of the load now the bottom. The chain hoist and/or both ratchet hoists can then be operated as necessary to lower the load to the floor. Be sure that proper blocking is provided under the load when appropriate.

THE CHAIN FALL ABOVE CAN BE OPERATED TO RAISE THE ENTIRE LOAD OFF THE FLOOR. THE RATCHET HOIST ON THE RIGHT IS THEN OPERATED TO RAISE THE END AND BRING THE LOAD TO A NEAR-VERTICAL POSITION. THIS IS DONE WITH THE LOAD COMPLETELY OFF THE FLOOR.

Figure 28 Raising the load off the floor, and then raising one end to begin the inversion.

1.5.0 Transferring a Load Between Hoists

There are times when loads may need to be moved horizontally, but must be raised above the floor to do so. This may be required due to obstructions on the floor, or perhaps to pass equipment across an open shaft. This requires two or more hoists, depending on how many transfers are needed to reach the destination. Moving a load with hoists and transferring the load weight from one hoist to another is often referred to as **drifting** a load.

As mentioned previously, since two or more hoists are involved, the safest approach is for each hoist to be operated by a single individual, while one qualified rigger supervises and directs their movements.

The process begins with attaching the load to a hoist and raising it off the floor (*Figure 30*). A second hoist, which will eventually assume the full load weight, is then attached to the same point. As the hoist on the left is raised, the load will be pulled to the left. It will also have a natural

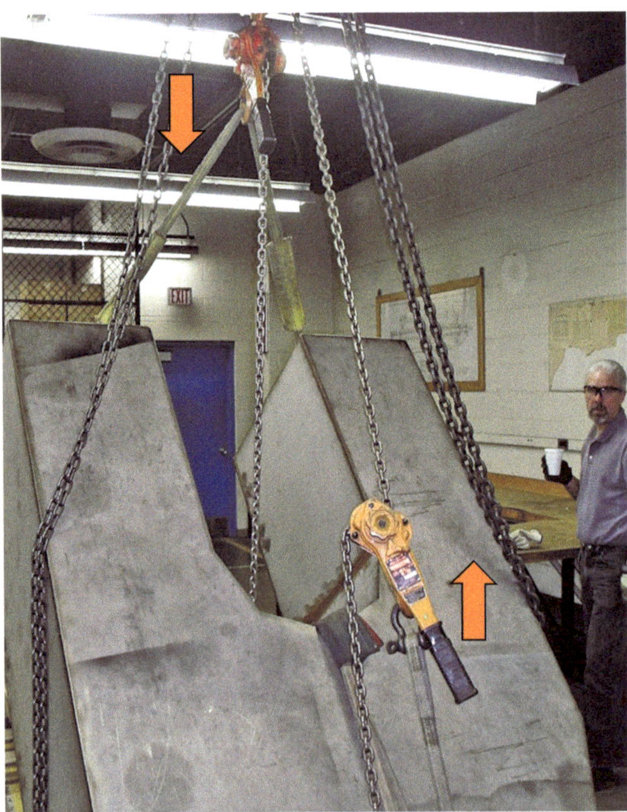

ONCE THE LOAD IS VERTICAL, SUSPENDED FROM THE TWO ATTACHMENT POINTS AT THE TOP, THE LOWER RATCHET HOIST IS DISCONNECTED AND MOVED TO THE OPPOSITE SIDE OF THE LOAD. IT IS THEN USED TO RAISE THAT END OF THE LOAD AS SHOWN HERE, RETURNING THE LOAD TO HORIZONTAL.

Figure 29 Completing the inversion.

NCCER – *Intermediate Rigger*

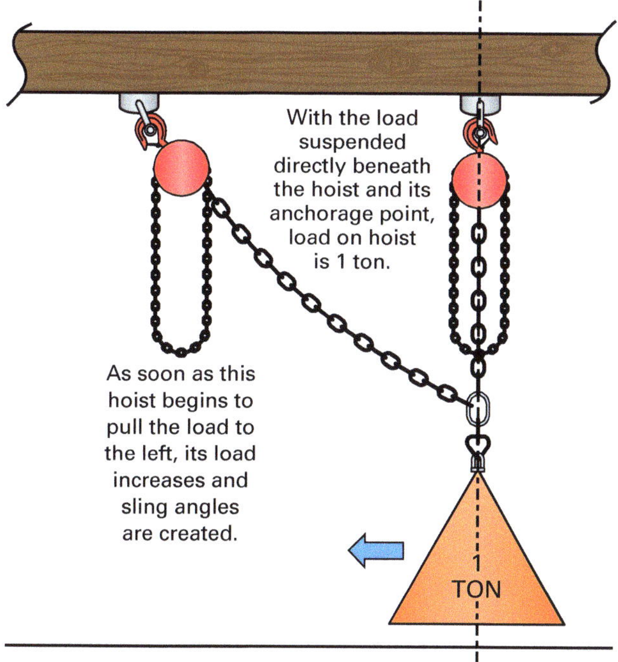

Figure 30 Preparing to drift a load.

tendency to rise. The hoist on the right must be lowered to compensate, allowing the load to move left. The operation of the two hoists must be coordinated to complete a smooth transfer.

As long as the load is hanging directly beneath the first hoist and its anchorage point, the hoist is supporting the entire load of 1 ton. The second hoist is merely connected. As soon as the second hoist is raised and pulls the load away from its perpendicular position, sling angles and the related loads are created. Because of the sling angles, the sum of the load imposed on both hoists will be greater than the load weight throughout the process. Each movement of the load to the left, by raising one hoist and lowering the other, results in new sling angles and a change in the load imposed on the slings and hoists. This continues until the load is directly beneath the second hoist, and the first hoist line goes slack when lowered further. At this point, the load on the second (re-

> **CAUTION**
>
> The weight of the rigging components must be accounted for in the load weight. For the sake of clarity, the weight of the rigging components is not considered here.

ceiving) hoist returns to 1 ton.

Sling angles in this application have the same effect as they do in other rigging applications.

The rules are also the same—sling angles less than 30 degrees to horizontal must be avoided due to the excessive stress. If the load is lifted off the floor a short distance, then drifted while maintaining the same relative elevation throughout, the smallest sling angles are formed at the beginning and at the end of the operation.

Refer to *Figure 31*. The diagram on the left side (A) shows that the hoists are anchored 12 feet (3.66 m) apart, with 10 feet (3.05 m) between the anchorage and the point of connection, forming a sling angle of roughly 40 degrees. Although not ideal, this sling angle is acceptable. As the load progresses from right to left, this sling's angle increases (improves). The other sling angle, however, which begins at 90 degrees, decreases as the load moves. Once the load is suspended under the left hoist, the sling angle to the right will have decreased to 40 degrees. Assuming the hoists, slings, hardware, and anchorage point are up to the task for the loads that will be imposed, this drifting operation can be accomplished without difficulty.

The diagram on the right side (B) shows the effect of raising the load higher. This shortens the vertical distance between the anchorage and point of the load connection. With the same 12-foot distance between the hoists, an unacceptable sling angle of less than 30 degrees results.

Several important points related to drifting operations are described as follows:

- Regardless of the specific dimensions, if the distance between the anchorage and the load connection is half or less than the distance between the two hoists, the sling angle will be less than 30 degrees.
- When the sling angle is less than 30 degrees, and no additional vertical distance is available, you must shorten the distance between the hoists. In other words, you must drift the load a shorter distance.
- Height above is an advantage when attempting to drift a load a significant distance. More vertical distance allows longer drifts without encountering unacceptable sling angles. It is best to keep loads as low to the floor as possible to maximize the distance above. Keeping the load as low as possible is the safest approach for all such tasks.
- As is the case for all similar operations, it is essential that anchorage points have the required capacity to support the weight of the load, rigging, and stress imposed by sling angles.
- Use softeners and padding as necessary to protect both the slings and the load surface, when appropriate.

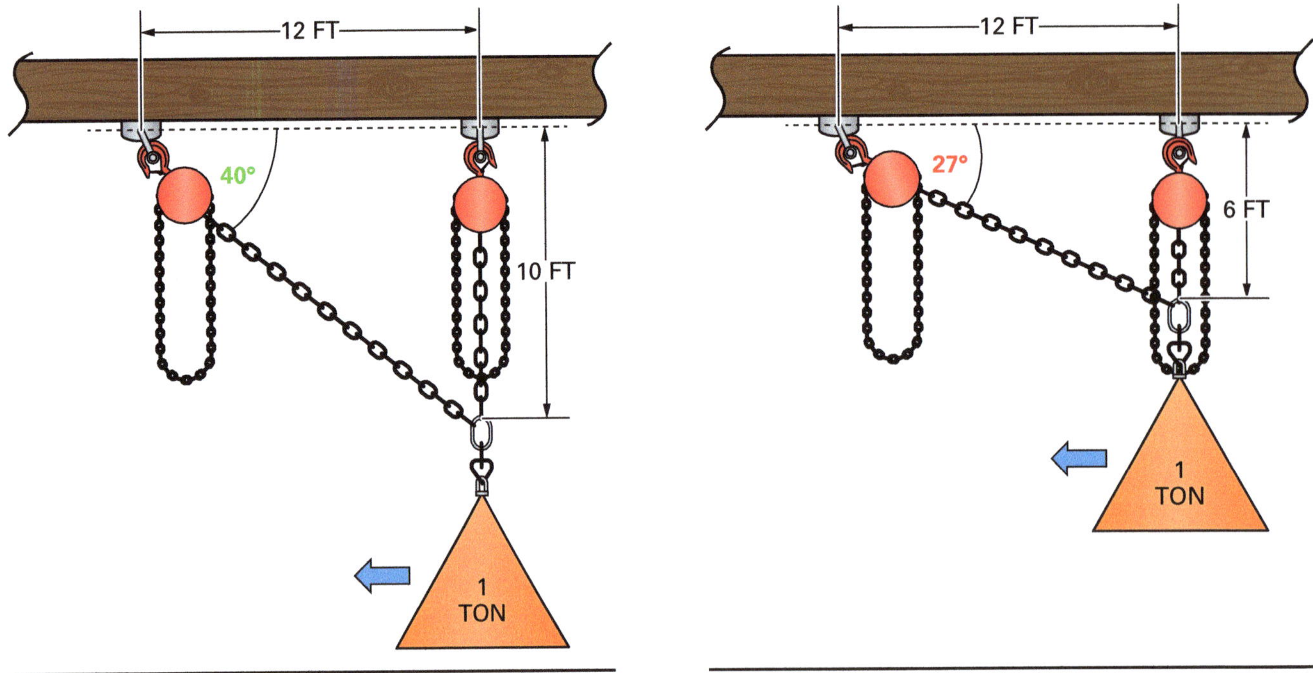

Figure 31 Initial sling angles for a drifting task.

Determining Sling Angle and Length

With some paper and a protractor, you can use a few site measurements to quickly confirm a sling angle. In *Figure 31*, the diagram labelled (A) shows that the distance between the points of connection is 12 feet. The vertical distance is 10 feet. Using the sling as the third side, sketch a triangle to scale on paper. Use any scale that is convenient, such as ½" = 1 foot. Graph paper makes it easy—you can simply count the boxes. Then, lay the protractor on the horizontal line and the vertex of the angle and read the angle.

How long is that sling assembly, from end to end? That's the hypotenuse of a right triangle. With the length of the two sides, you can determine the length of the hypotenuse using this formula:

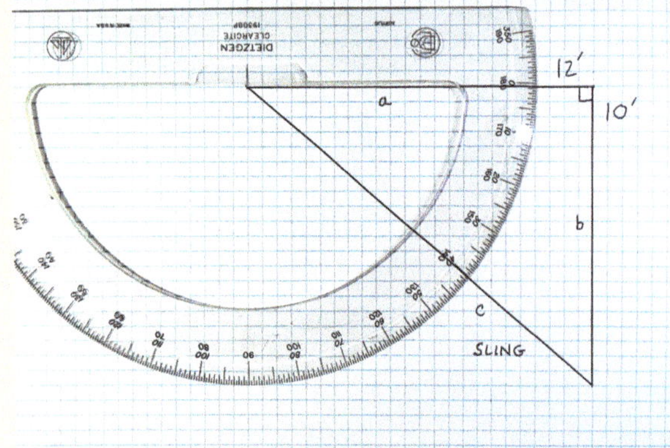

$$a^2 + b^2 = c^2$$
$$12^2 + 10^2 = c^2$$
$$144 + 100 = c^2$$
$$244 = c^2$$
$$\sqrt{244} = \sqrt{c^2}$$
$$15.62 \text{ ft} = c$$

NCCER – *Intermediate Rigger*

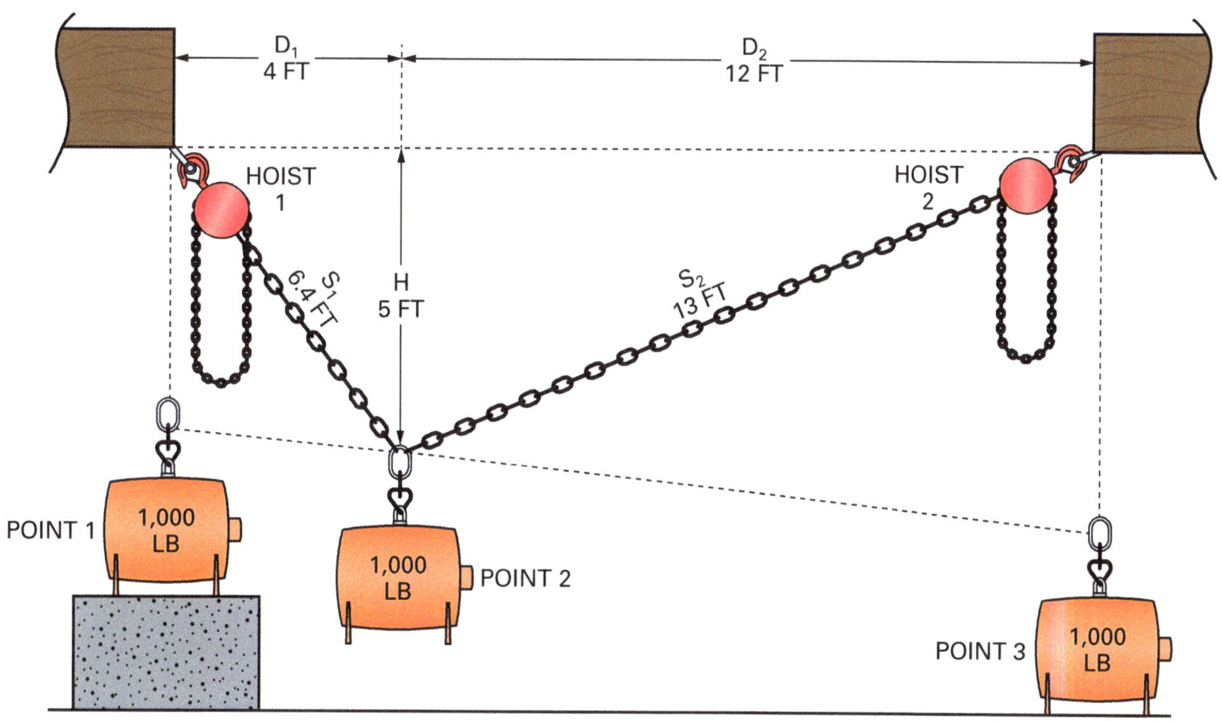

TENSION IN SLINGS S$_1$ AND S$_2$ – POINT 2

TENSION IN SLING S$_1$ = T$_1$

$$T_1 = \frac{LOAD \times (D_2 \times S_1)}{H \times (D_1 + D_2)}$$

$$T_1 = \frac{1{,}000 \times (12 \times 6.4)}{5 \times (4 + 12)}$$

$$T_1 = \frac{1{,}000 \times 76.8}{5 \times 16}$$

$$T_1 = \frac{76{,}800}{80}$$

$$T_1 = 960 \text{ LB}$$

TENSION IN SLING S$_2$ = T$_2$

$$T_2 = \frac{LOAD \times (D_1 \times S_2)}{H \times (D_1 + D_2)}$$

$$T_2 = \frac{1{,}000 \times (4 \times 13)}{5 \times (4 + 12)}$$

$$T_2 = \frac{1{,}000 \times 52}{5 \times 16}$$

$$T_2 = \frac{52{,}000}{80}$$

$$T_2 = 650 \text{ LB}$$

TOTAL TENSION ON SLINGS = 1,610 LB

NOTE: Equations only valid when the anchorage points are at the same elevation.

Figure 32 Hoist-to-hoist sling load calculations.

1.5.1 *Calculating Sling Tension Throughout the Drift*

To ensure a safe lift and select the appropriate rigging components, sling tension needs to be checked at various points throughout the drift. As described earlier, the full weight of the load will progress from one hoist to another. At all points between, the sum of the tension on both slings will be something greater than the load weight alone.

In *Figure 32*, a 1,000-lb load is being drifted 16 feet (4.88 m). At Points 1 and 3, there are no sling angles to create additional tension, and the weight on the hoist above is 1,000 lb. At all points between, sling angles will add tension to the total load on the hoists. The steps below describe the sling load calculations for Point 2.

The two slings are labeled as S_1 and S_2. The loads on each sling are identified as T_1 and T_2. As shown in *Figure 32*, the tension on sling S_1 (T_1) is found using the following equation:

$$T_1 = \frac{\text{load} \times (D_2 \times S_1)}{H \times (D_1 + D_2)}$$

The tension on sling S_2 (T_2) is found with a similar equation. Note that the denominators of the two equations are the same:

$$T_2 = \frac{\text{load} \times (D_1 \times S_2)}{H \times (D_1 + D_2)}$$

At Point 2, the sling on the left, carrying 960 lb, is still carrying most of the load weight. It's sling angle is greater than that of the right sling, and therefore less additional tension is being applied to it. Of the 650-lb tension applied to the right sling, a larger portion of its load can be attributed to stress from the reduced sling angle. As the load progresses from one point to another, the proportions of load weight and sling stress will continually change.

Sling load calculations may need to be done for several points along the path of a drift plan. This is necessary when the load will also be lifted to various heights, decreasing the sling angle. With practice, riggers will be able to envision where the sling angles will be at their lowest along the load's route. Remember that sling tension is at its highest where the angle is at its lowest.

Additional Resources

Bob's Rigging and Crane Handbook. Current edition. Leawood, KS: Pellow Engineering Services.

Rigging Handbook. Current edition. Jerry A. Klinke. Stevensville, MI: ACRA Enterprises, Inc.

Willy's Signal Person & Master Rigger Handbook. Current edition. Ted Blanton Sr., Robert O'Leary, Joe Crispell, and Ted Blanton Jr. Lake Mary, FL: NorAm Productions, Inc.

1.0.0 Section Review

1. A braided wire rope sling used in a choker hitch has a vertical hitch WLL of 1,000 lb. Determine its WLL at a choke angle of 100 degrees.

   a. 1,000 lb
   b. 870 lb
   c. 740 lb
   d. 620 lb

2. When skidding a 3,500-pound concrete runoff-holding tank on level oak skids, what would be a reasonable estimate for the line pull required to move the load over the skids?

   a. 1,775 lb
   b. 1,575 lb
   c. 1,375 lb
   d. 1,075 lb

3. When there is an even number of parts of line to be reeved on a block and tackle, the end of the rope will terminate at the _____.

   a. upper block
   b. lower block
   c. load
   d. anchor point

4. Considering where the CG is for a load to be inverted using a single hoist helps to determine _____.

   a. the weight of the load
   b. the sling angle
   c. where to attach the hoist to the load
   d. which side is the bottom

5. When planning to drift a load between two hoists, if the distance between the anchorage and the load connection is half or less than the distance between the two hoists, the _____.

   a. hoists will need to be moved farther apart
   b. anchorage points will need to be lowered
   c. sling angle will be more than 30 degrees
   d. sling angle will be less than 30 degrees

SUMMARY

Some form of sling is used to secure the load in nearly every lift. The ability to choose the correct sling and use it in the prescribed manner are essential skills that every rigger must master. It is critically important to know how to derate the sling capacity based on the sling configuration. It is equally important to understand how sling angles increase tension beyond the weight of the load itself.

Riggers are often required to handle loads indoors, where the convenience of a crane isn't an option. Loads of all types must be moved into place indoors. There are a number of devices used to roll or skid heavy loads into place across a floor. In other cases, the load must be moved through the air using block and tackle, ratchet hoists, or chain hoists. A load may also need to be turned or inverted. These unique rigging tasks require attention to detail and an understanding of the stress imposed on each of the elements involved. Riggers need to master details such as these in order to be successful in their profession.

1. The length of a sling is the ______.
   a. overall length of the sling from one end to the other
   b. distance from the center of the eye at one end to the center of the eye at the other end
   c. distance between the loadbearing points at each end
   d. length of the main part of the sling, not including the end fittings

2. What precaution must a rigger observe when using the dual-sling bridle formula WLL = 2W ÷ SAF?
   a. Both slings must be of the same material.
   b. Both slings must support the load equally.
   c. The formula should only be applied to wire rope.
   d. The formula is valid only when both legs are vertical hitches.

3. Which of the listed items would a rigger use mainly for vertical positioning of a heavy piece of equipment?
   a. Mechanical rollers
   b. Grip hoist
   c. Spreader jack
   d. Bottle jack

4. Which equipment used to move a load laterally over a level concrete surface should require the smallest line pull?
   a. Rollers
   b. Skids
   c. Grip hoists
   d. Pry bars

5. On a block and tackle, the line you pull to lift an object is referred to as the ______.
   a. secondary line
   b. hauling line
   c. primary line
   d. scuttle line

6. On a block and tackle, what is the line pull required to lift an object weighing 400 lb with three parts of line reeved? Round your answer to the nearest whole number.
   a. 400 lb
   b. 300 lb
   c. 266 lb
   d. 134 lb

7. When inverting a load on its side with a single hoist, always lift from an anchorage point ______.
   a. above the load CG
   b. below the load CG
   c. even with the load CG
   d. as close as possible to the CG

8. Which of the following is a *true* statement about inverting a load with two hoists?
   a. Sling stress is never a concern in this task.
   b. The points of connection to the load should always be well below the load CG.
   c. At one point, a single hoist will need to carry the load.
   d. Chain hoists cannot be used for this task.

9. When drifting a load and maintaining the same relative elevation throughout the task, the smallest sling angles are formed by either hoist at ______.
   a. the beginning of the drift
   b. the end of the drift
   c. both the beginning and the end
   d. the midpoint

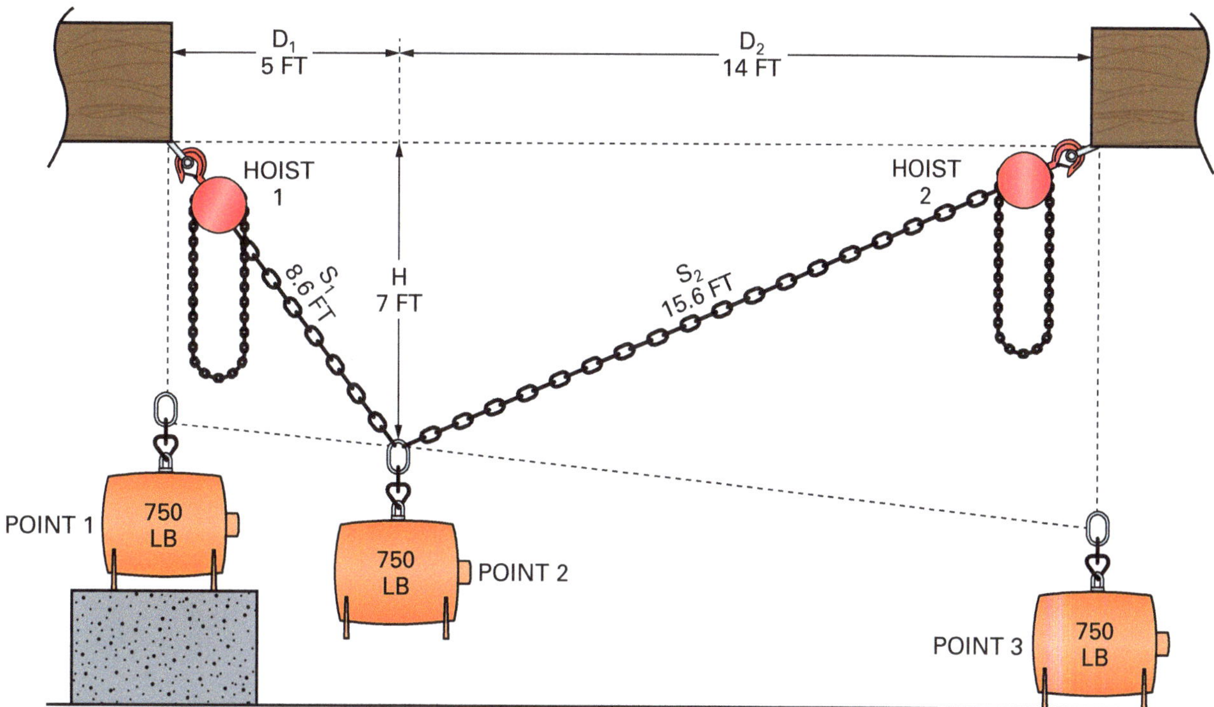

Determine sling tension using the following equation:

$$T_1 = \frac{\text{load} \times (D_2 \times S_1)}{H \times (D_1 + D_2)}$$

Figure RQ01

10. Refer to *Figure RQ01*. Determine the tension on Sling S_1 (assigned the variable of T_1) at Point 2, using the provided equation. Round your answer up to the nearest whole number.

    a. 803 lb
    b. 750 lb
    c. 679 lb
    d. 581 lb

Trade Terms Introduced in This Module

Anchorage point: A point used to fix the position of a block or other type of hoist or pulling device. An anchorage point must be able to easily handle the load weights and line pulls associated with any sling angle imposed on it, without fear of failure.

Center of gravity (CG): The point where an object's mass, and therefore its weight, is concentrated. The concept is useful for determining stability, the balance point, and leverage.

Choke angle: The outside angle between the vertical part of a choker hitch and the part of the sling passing through the eye, hook, or shackle forming the choke.

Coefficient of friction (CF): A ratio that expresses a comparison between the force necessary to move an object over the surface of another material, and the pressure between the two materials.

D/d ratio: For rope-like slings used in hitches that wrap around the load, the ratio of the diameter of the load to the diameter of the rope. Manufacturers establish minimum D/d ratios to avoid overstressing the sling with a bend that is too sharp.

Derate: The process of reducing the working load limit of a sling according to standard rules to account for the large stresses that exist in a sling when used in certain applications.

Drifting: The process of moving a load laterally through the air, using multiple hoists and transferring the load weight from one hoist to another.

Hitch: Any method of attaching a sling to a load. Common hitches include vertical, bridle, basket, and choker.

Invert: To move a load from the horizontal position, for example, to any other position, such as over on its side, vertical on its end, or fully inverted to lie on its top.

Sling angle: The angle formed between the sling and the horizontal when under tension.

Sling-angle factor (SAF): A factor that permits calculating the tension in a sling supporting a known load at an angle. Equal to the length of the sling between its bearing points divided by the vertical distance between those points (SAF = L ÷ H).

Tuggers: Motorized or engine-powered winches designed to replace manpower to pull the rope used for lifting or moving equipment.

Working load limit (WLL): The weight capacity a manufacturer certifies a rigging component can lift under stated conditions. The standard sling WLL is provided for a vertical hitch. Riggers will typically apply standard rules to derate WLLs when slings are used in other configurations.

Additional Resources

This module is intended as a thorough resource for task training. The following reference materials are recommended for further study.

Bob's Rigging and Crane Handbook. Current edition. Leawood, KS: Pellow Engineering Services.
Rigging Handbook. Current edition. Jerry A. Klinke. Stevensville, MI: ACRA Enterprises, Inc.
Willy's Signal Person & Master Rigger Handbook. Current edition. Ted Blanton Sr., Robert O'Leary, Joe Crispell, and Ted Blanton Jr. Lake Mary, FL: NorAm Productions, Inc.

Figure Credits

Section Review Answer Key

Answer Section One	Section Reference	Objective
1. b	1.1.4; Figure 10	1a
2. b	1.2.3; Table 3	1b
3. a	1.3.2; Figure 24	1c
4. c	1.4.1	1d
5. d	1.5.0	1e

1.0.0 SECTION REVIEW

Question 1

The choke angle at 100 degrees reduces the WLL to 87 percent of the sling's base capacity. Therefore, calculate 87 percent of the base capacity to determine the WLL:

$0.87 \times 1{,}000 \text{ lb} = 870 \text{ lb}$

The WLL is **870 lb**.

Question 2

Concrete on wooden skids has a coefficient of friction (CF) of 0.45. Therefore:

$0.45 \times 3{,}500 \text{ lb} = 1{,}575 \text{ lb}$

The required line pull is **1,575 lb**.

NCCER CURRICULA — USER UPDATE

NCCER makes every effort to keep its textbooks up-to-date and free of technical errors. We appreciate your help in this process. If you find an error, a typographical mistake, or an inaccuracy in NCCER's curricula, please fill out this form (or a photocopy), or complete the online form at **www.nccer.org/olf**. Be sure to include the exact module ID number, page number, a detailed description, and your recommended correction. Your input will be brought to the attention of the Authoring Team. Thank you for your assistance.

Instructors – If you have an idea for improving this textbook, or have found that additional materials were necessary to teach this module effectively, please let us know so that we may present your suggestions to the Authoring Team.

NCCER Product Development and Revision
13614 Progress Blvd., Alachua, FL 32615

Email: curriculum@nccer.org
Online: www.nccer.org/olf

❏ Trainee Guide   ❏ Lesson Plans   ❏ Exam   ❏ PowerPoints   Other _______________

Craft / Level: _______________________________________   Copyright Date: _______________

Module ID Number / Title: ____

Section Number(s): ____

Description: ____

____

____

____

Recommended Correction: ____

____

____

____

Your Name: ____

Address: ____

____

Email: ____   Phone: _______________

Load Dynamics

OVERVIEW

Mobile cranes handle objects that are heavy, bulky, and difficult to move by any other means. Crane structures can be complex, and their configurations change from job to job depending on the task at hand. However, all cranes rely on the basic principle of a lever to provide a stable platform for lifting a load. The term *load dynamics* encompasses this simple principle as well as other factors that affect leverage as a lift progresses. This module presents the principles of load dynamics that permit a crane to safely accomplish its tasks.

Module 21206

Trainees with successful module completions may be eligible for credentialing through the NCCER Registry. To learn more, go to **www.nccer.org** or contact us at 1.888.622.3720. Our website has information on the latest product releases and training, as well as online versions of our *Cornerstone* magazine and Pearson's product catalog.

Your feedback is welcome. You may email your comments to **curriculum@nccer.org**, send general comments and inquiries to **info@nccer.org**, or fill in the User Update form at the back of this module.

This information is general in nature and intended for training purposes only. Actual performance of activities described in this manual requires compliance with all applicable operating, service, maintenance, and safety procedures under the direction of qualified personnel. References in this manual to patented or proprietary devices do not constitute a recommendation of their use.

21206 V3

From *Intermediate Rigger, Trainee Guide*, NCCER.

Objectives

When you have completed this module, you will be able to do the following:

1. Explain how the concepts of leverage and stability are related to mobile crane operation.
   a. Explain how rotational forces affect leverage and how they can be balanced.
   b. Identify the primary factors related to crane stability.
   c. Identify other significant factors that affect crane stability.

Performance Tasks

This is a knowledge-based module; there are no Performance Tasks.

Trade Terms

Acceleration
Buoyancy
Center of gravity (CG)
Center of rotation
Crane moment
Dynamics
Effective weight
Fulcrum
Impact loading
Law of moments
Lever

Leverage
Load moment
Load moment indicator (LMI)
Load radius
Moment
Moment arm
Momentum
Pendulum effect
Stability
Tipping axis

Industry Recognized Credentials

If you are training through an NCCER-accredited sponsor, you may be eligible for credentials from NCCER's Registry. The ID number for this module is 21206. Note that this module may have been used in other NCCER curricula and may apply to other level completions. Contact NCCER's Registry at 888.622.3720 or go to **www.nccer.org** for more information.

Contents

Figures and Tables

1.0.0 MOMENTS AND STABILITY

Objective

Explain how the concepts of leverage and stability are related to mobile crane operation.
   a. Explain how rotational forces affect leverage and how they can be balanced.
   b. Identify the primary factors related to crane stability.
   c. Identify other significant factors that affect crane stability.

Trade Terms

Acceleration: In common usage, the rate of increase of an object's speed per second. The rate of decrease of speed is commonly called .

Buoyancy: The upward force exerted by a liquid or gas (both fluids) on an object immersed in the fluid.

Center of gravity (CG): The point at which the entire weight of an object is concentrated, such that supporting the object above this specific point would result in its remaining balanced in position.

Center of rotation: In crane operations, the vertical axis through the center of the swing circle around which the upperworks rotates.

Crane moment: The product of the entire crane's center of gravity and its distance from the tipping axis along the lever arm toward the load's center of gravity. Commonly called the crane's *leverage* or *lever force*.

Dynamics: A general term that refers to forces, their effects, and how they change from one moment to the next.

Effective weight: Also referred to as the apparent weight. The weight of a load plus or minus the effect of other factors, such as acceleration, wind, and buoyancy. For example, a load that accelerates while falling exerts a force that is greater than the actual weight of the load when it is suddenly stopped.

Fulcrum: The point of support on which a lever pivots.

Impact loading: A short-duration force that is suddenly exerted when a moving object is slowed or stopped.

Law of moments: A scientific principle that defines the conditions under which a lever-and-fulcrum system is balanced and stable.

Lever: A simple machine consisting of a rigid arm pivoting on a fulcrum. Force can be transferred from one end to the other as it pivots, changing the force's direction and size, or two forces acting on its ends can work against each other.

Leverage: A common term describing the action of a lever, especially for prying or lifting a heavy object. See *moment*.

Load moment: The force applied to the crane by the load; the leverage of the load, opposing the leverage of the crane. The load moment is calculated by multiplying the gross load weight by the horizontal distance from the tipping fulcrum to the center of gravity of the suspended load. The load moment is usually reported to the operator as a percentage of the crane's capacity at the present set of conditions. As those conditions change, such as the boom angle, the load moment changes as well.

Load moment indicator (LMI): A system that aids the equipment operator by directly or indirectly sensing the overturning moment on the equipment. It compares the lifting conditions to the equipment's rated capacity, and when the rated capacity is reached, it disables equipment functions that can increase the severity of loading on the equipment, such as hoisting or telescoping the boom further out.

Load radius: The horizontal distance between the crane's center of rotation and the vertical hoist line or tackle with the load applied. On load charts, the load radius is determined by the boom angle and its length to the boom tip.

Moment: The effect of a force acting perpendicularly on a lever at a certain distance from its fulcrum. Commonly called *leverage*. Mathematically, moment is equal to the product of the force and its moment arm.

Moment arm: In a lever system, the distance between the point where a force is applied to a lever and the lever's fulcrum.

Momentum: A property of all moving objects that is equal to the product of their mass and velocity. In equation form, $M = m \times v = mv$, where M is the object's momentum, m is its mass, and v is its velocity.

Pendulum effect: The variable force exerted by a swinging load on the crane structure along the hoist line, like the swing of a pendulum. The load's excess force is maximum at the bottom of the swing.

A mobile crane is a machine that lifts and moves loads within its designed capabilities. Cranes apply a mechanical advantage to a load to produce the desired results. The loads being moved by a crane exert forces upon the machinery that are predictable. In order to understand a crane's behavior, you must gain a working knowledge of how forces act on a crane and the factors that contribute to stable and unstable conditions.

1.1.0 Leverage and Balance

Crane operators need to understand how forces acting on an object can both cause and prevent rotation around a fixed point. Crane **stability** relates to preventing tipping and taking steps to ensure that the crane is always in a more powerful position than its load.

1.1.1 Levers and Moments

A crane is a practical example of the application of the **lever** principle that you probably learned about in grade school. The two physical pieces of a lever are the lever arm and a **fulcrum** around which it can pivot (*Figure 1*). A lever is a simple machine that takes a force exerted on one end of the lever to exert a force on an object at the other end as it pivots around its fulcrum. Think of a claw hammer or a pry bar; both these tools use the lever principle.

Consider a lever from a more technical viewpoint. There are some measurable dimensions that affect the way a lever works. Refer to *Figure 2*. A downward force A (F_A) is exerted at the left end of the lever arm. The point where the force is applied is a distance D_A from the fulcrum. At the other end, force B (F_B) is exerted downward at a distance D_B from the fulcrum. When using a hammer or a pry bar, the farther from the fulcrum that you exert a force, the more effect it seems to have. For a lever, the product of a force exerted on the lever and the distance of that force from the fulcrum is called the **moment** of the force. The moment combines the effects of

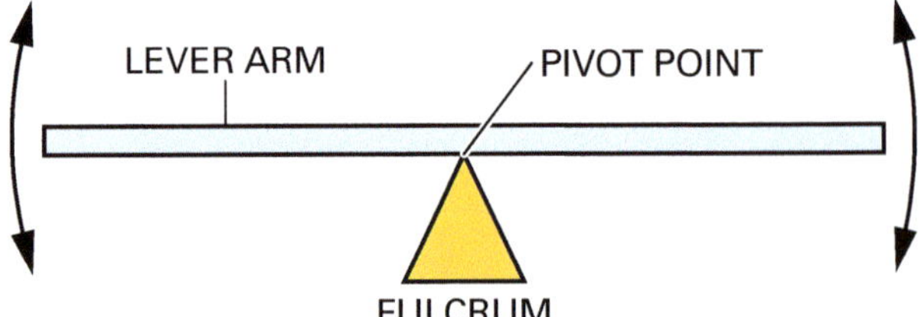

Figure 1 A simple lever.

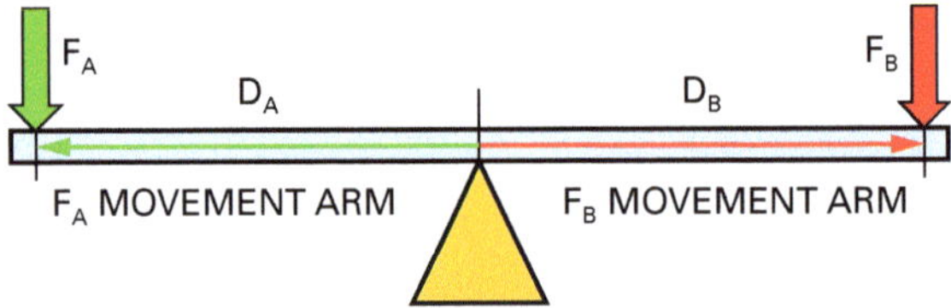

Figure 2 A lever and its force moments.

force and distance into a single unit of measure. When discussing levers, the moment of a force is commonly called its **leverage**. The American Society of Mechanical Engineers (ASME) refers to moments in *ASME Standard B30.5, Mobile and Locomotive Cranes*, so the term *moment* will be used in this module.

The moment of a force is a measurable quantity with its own units. Since a force's moment is the product of a distance and a force, its unit in the Imperial system is the foot-pound (ft·lb). (The dot in the unit represents multiplication.) In the metric system, a force's moment has the unit newton-meter (N·m). The perpendicular distance from the fulcrum to the exerted force is called the force's **moment arm** or lever arm.

Refer again to *Figure 2*. You can see that F_A would cause the lever to rotate in the counterclockwise direction around the fulcrum and F_B would cause it to rotate clockwise. If the product of F_A and D_A (the moment of F_A) is greater than $F_B \times D_B$ (the moment of F_B), then the rotation is counterclockwise; if the moment of F_B is greater than that of F_A, then the motion is clockwise. So another important property of a moment is that it has direction around a point or axis of rotation. This idea is behind the **law of moments**, which can be stated in the following equation:

$$F_A \times D_A = F_B \times D_B$$

This equation is true only if the two moments act around the fulcrum in opposite directions and no rotational motion results; there is balance. Note that if any of the factors change, such as the force becoming greater on one side, then another factor must change for balance to continue. On the other hand, if one of the factors changes and all others remain the same, rotation will occur until a new position of balance is found. If the change is too

Moments Like These

The idea that describes a force acting on a lever around a pivot—*moment*—is believed to originated with the Greek philosopher Archimedes, who supposedly said, "Give me a place to stand, and I shall move the world," referring to his discovery of the usefulness of the lever.

The word *moment* itself seems to have come from the Latin word, *momentorum*, created by a translator of Archimedes' writings in 1565 to describe his principle of leverage. Today, physicists use the word to describe many properties of moving objects revolving or rotating around a center of motion.

great, then balance cannot be attained and one side will fall toward the ground. You will see later in this module how the law of moments helps describe and predict crane stability.

1.1.2 Leverage and Stability

Up to this point, the lever and the forces acting on it have been viewed as an isolated system operating without regard for the surroundings. Real lever systems have a fulcrum fixed in place and are generally limited in how far they can move, like a seesaw at a playground. Imagine two children playing on a seesaw (*Figure 3*). If one child is heavier than the other, the end of the seesaw with the heavier child will sit on the ground. The seesaw can't rotate any further because the ground

is exerting an upward force on that end to stop its motion, even though the moments of the children's weights are not equal.

> **NOTE**
>
> Recall that weight is a force exerted by gravity on an object in proportion to its mass. The more mass an object has, the more it weighs.

What is the size of the force exerted by the ground? Study the following example:

A 100-pound (446 N) girl and her 80-pound (357 N) brother are riding a seesaw at the playground. They each sit 4 feet (1.2 m) from the fulcrum. What is the difference in moments for their two weights?

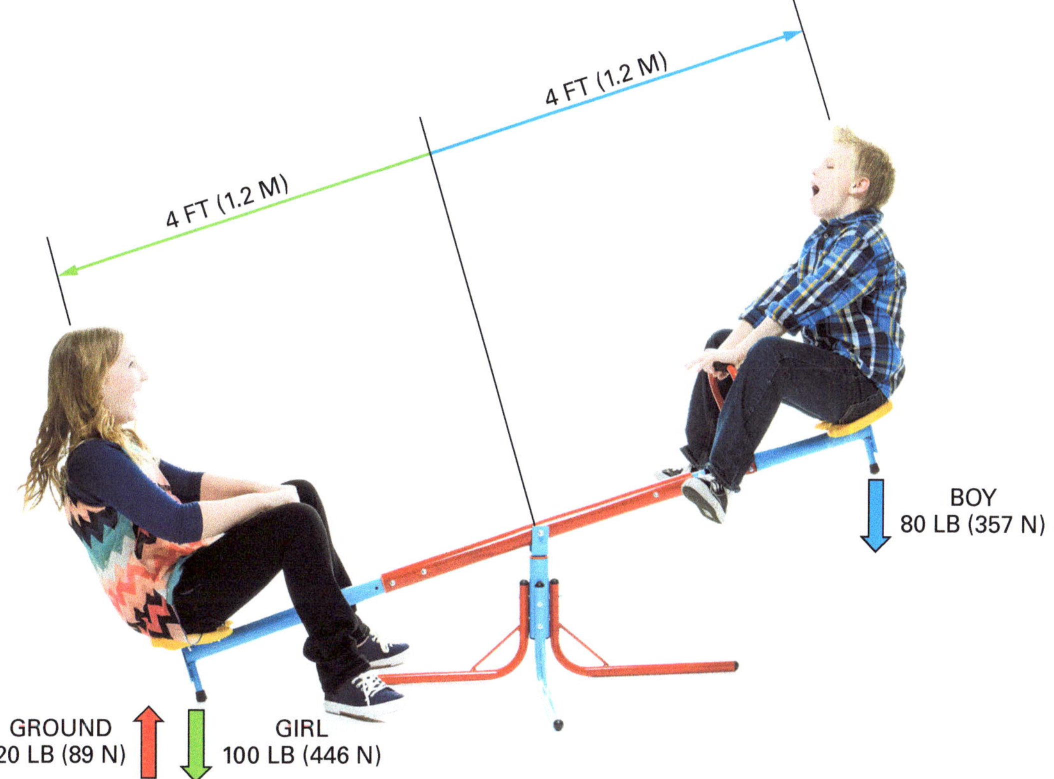

Figure 3 Seesaw lever example.

Calculate the girl's and boy's moments using the following formula:

$$\text{Moment} = \text{moment arm} \times \text{force}$$

Girl's moment:

Girl's moment = girl's moment arm × girl's weight
Girl's moment = 4 ft × 100 lbs
Girl's moment = 400 ft·lbs (542 N·m)

Boy's moment:

Boy's moment = boy's moment arm × boy's weight
Boy's moment = 4 ft × 80 lbs
Boy's moment = 320 ft·lbs (434 N·m)

Difference of moments:

$$400 \text{ ft·lbs} - 320 \text{ ft·lbs} = 80 \text{ ft·lbs} (108 \text{ N·m})$$

Since the girl is heavier than the boy, this difference moment is acting downward toward the ground at a distance of 4 ft (1.2 m) from the seesaw fulcrum at the girl's end. Divide the difference moment by the moment arm to determine the downward force exerted on the ground:

$$80 \text{ ft·lbs} \div 4 \text{ ft} = 20 \text{ lbs} (89 \text{ N})$$

To keep the seesaw from moving, the ground exerts an equal but upward force on the seesaw at the end of the seesaw. In this condition, all forces and moments are equal and the seesaw is balanced or, better said, stable. More importantly, the ground can support any expected force the seesaw can exert on it, which contributes to the stability of the seesaw no matter how heavy the people are who ride it. Even if you add more weight to the girl's side of the seesaw, the seesaw doesn't move down any farther. Instead, the downward force exerted on the ground just increases. The firmness of the ground is another key concept to keep in mind when evaluating crane stability.

1.1.3 The Crane Stability Model

With a little imagination, the simple lever system of *Figure 2* can be transformed into a bent shape that sort of resembles a crane (*Figure 4*). On the left side of the fulcrum is the crane base, the counterweights, and the rest of the upperworks, which will be discussed in more detail shortly. The right side represents the remaining portion of the crane's upperworks, including the boom, jib, load block, headache ball, lifting rope, all load rigging, and the load.

Each of the circular symbols in *Figure 4* represents the center of gravity (CG) of the major crane components shown. Recall that the CG of an object is the point inside it where you can assume all the mass exists. In other words, its weight is concentrated at the CG. To make the crane stability model useful, you must combine the CGs of all the different parts of a crane into one, which becomes the crane's CG (*Figure 5*). The load's CG is centered inside the load, as shown.

In NCCER Module 21102, *Basic Principles of Cranes*, the fulcrum was referred to as the *tipping fulcrum*, which was fitting for operator trainees learning crane stability concepts as two-dimensional problems. However, actual cranes exist and operate in three dimensions, so the tipping fulcrum concept needs to be expanded.

As with the seesaw, a mobile crane is supported by the ground. Cranes contact the ground in various ways. Crawler cranes rest on their tracks. Pick-and-carry cranes, and occasionally rough-terrain (RT) and all-terrain (AT) cranes, rest on their wheels (on rubber) during lifts. All other types of wheeled cranes generally rest on outriggers unless engaged in travel operations. In three dimensions, the tipping fulcrum is more than a point on a two-dimensional figure; it is an imaginary line on the ground called the tipping axis.

The crane's tipping axis at any given moment during operation is determined by the shape and location of the areas where the crane is supported by the ground, and the location of the boom relative to them. For a crawler crane, the possible tipping axes are along both sides of the treads and at the front and back ends of the treads where they contact the ground. For on-rubber wheeled cranes, the tipping axes are lines along each side, and at the front and back wheels that pass through the patches where the wheels contact the ground. For cranes on outriggers, the tipping axes are lines passing through the centers of the outrigger floats on both sides, and at the front and back outriggers. *Figure 6* shows examples of the tipping axes for various cranes.

As you can see, a crane's tipping axis is an imaginary horizontal line on the ground around which an unstable crane could rotate and fall over. You might think of it as a hinge point.

Returning to the crane stability model, the crane's moment is calculated along an imaginary line passing through the crane's CG and the CG of the load. This line can be thought of as the crane-load lever arm. When looking at the plan view of the crane and load (*Figure 7*), the crane moment arm is the distance between its CG and the point where this line intersects the applicable tipping axis. The load moment arm is the distance of the load's CG from the tipping axis along the same line.

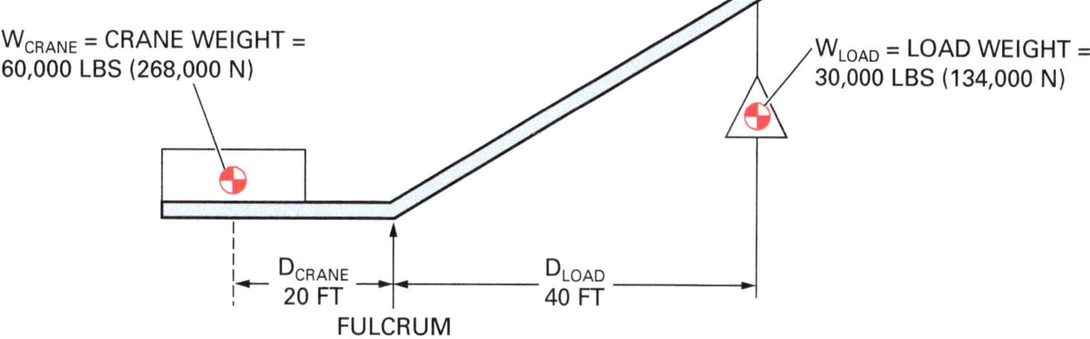

Figure 4 The crane stability model.

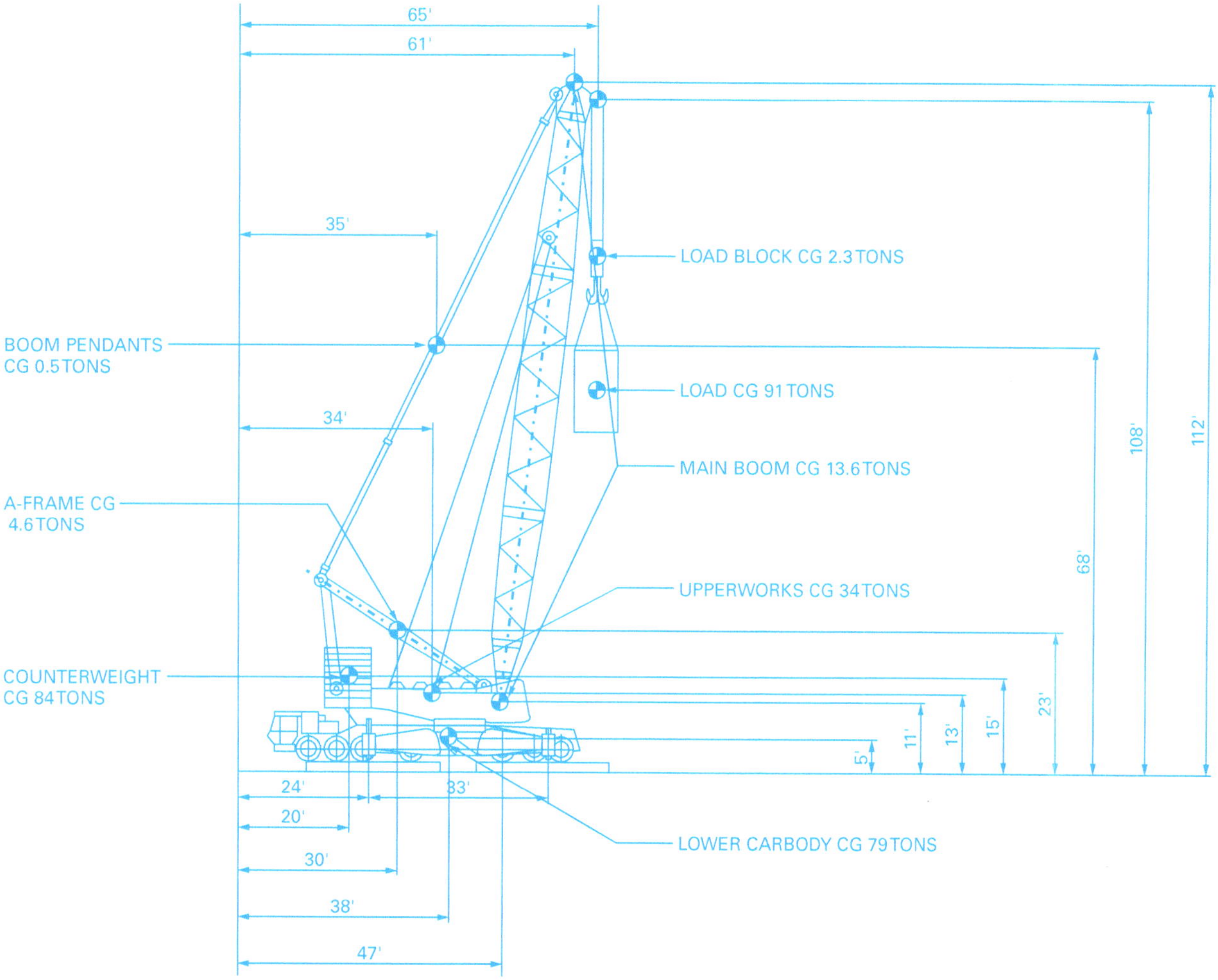

Figure 5 Crane component CGs affecting the crane's stability.

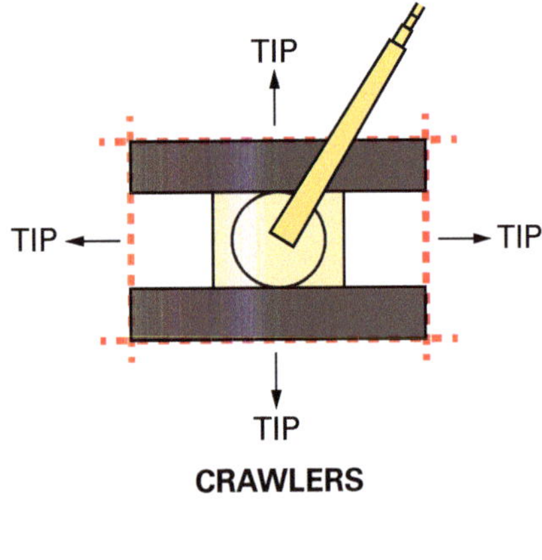

CRAWLERS

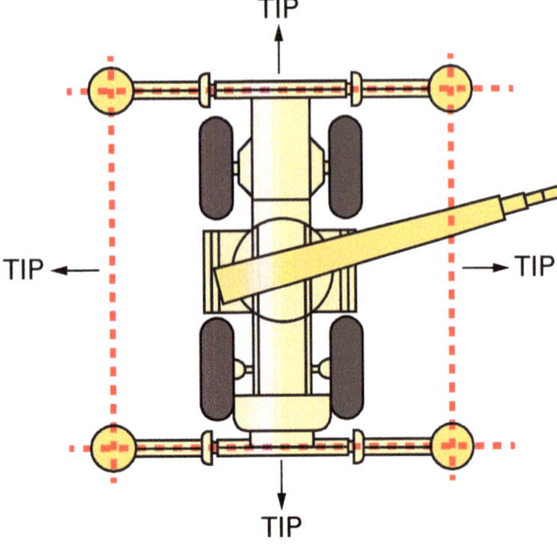

ON OUTRIGGERS

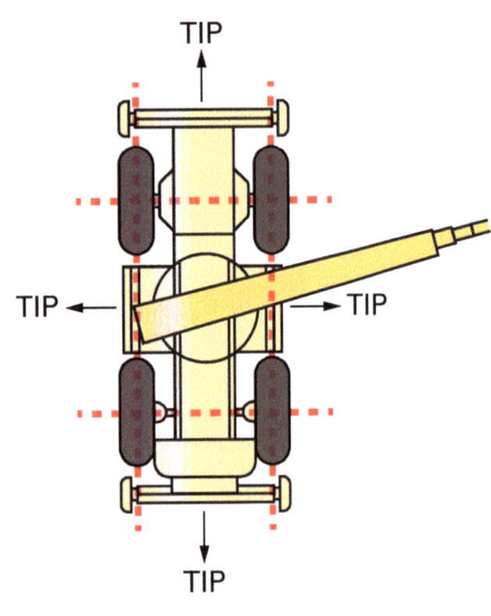

ON RUBBER

Figure 6 Identification of tipping axes for various crane types.

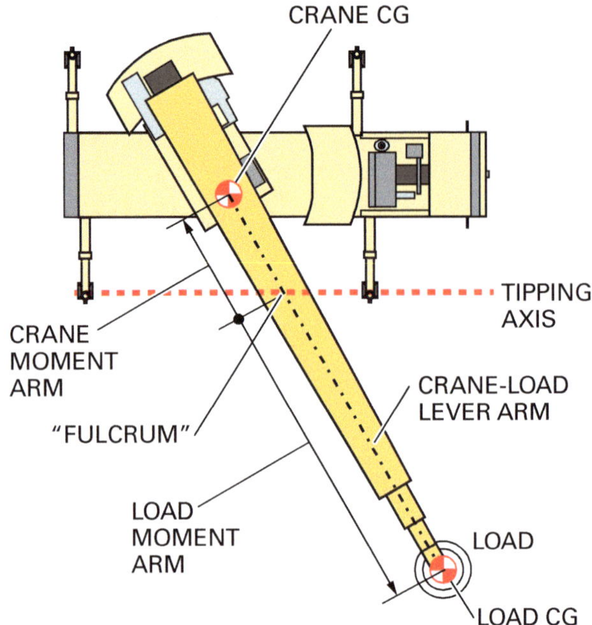

Figure 7 Establishing the crane and load moment arms.

The number of variables that are introduced into the calculation of moments make crane operations more complicated and dynamic than that of a simple seesaw. During a lift with a crane, the distance from the crane's CG to the tipping axis will change as the crane upperworks is rotated and the boom's position is changed (*Figure 8*). The distance of the load's CG from the tipping axis will also change during a swing and when the boom moves in or out. As the boom is loaded, it will also deflect, causing an additional change in the load moment arm. Movement causes the stability of the crane to be dynamic (constantly changing).

A mobile crane is stable when its moment relative to the tipping axis is greater—preferably much greater—than the load's moment. As with the seesaw example, the difference in moments is made up by the upward force of firm ground acting on the crane supports, which is essential for stable crane operations. A crane's stability decreases as the load's moment arm increases or as heavier loads are lifted at the same moment arm. If the load moment exceeds the crane's moment, the crane will tip. If both moments are equal, the crane's stability is in jeopardy. Theoretically, in an equal-moment situation (perfect balance), the crane would tip over when someone simply

NCCER – *Intermediate Rigger*

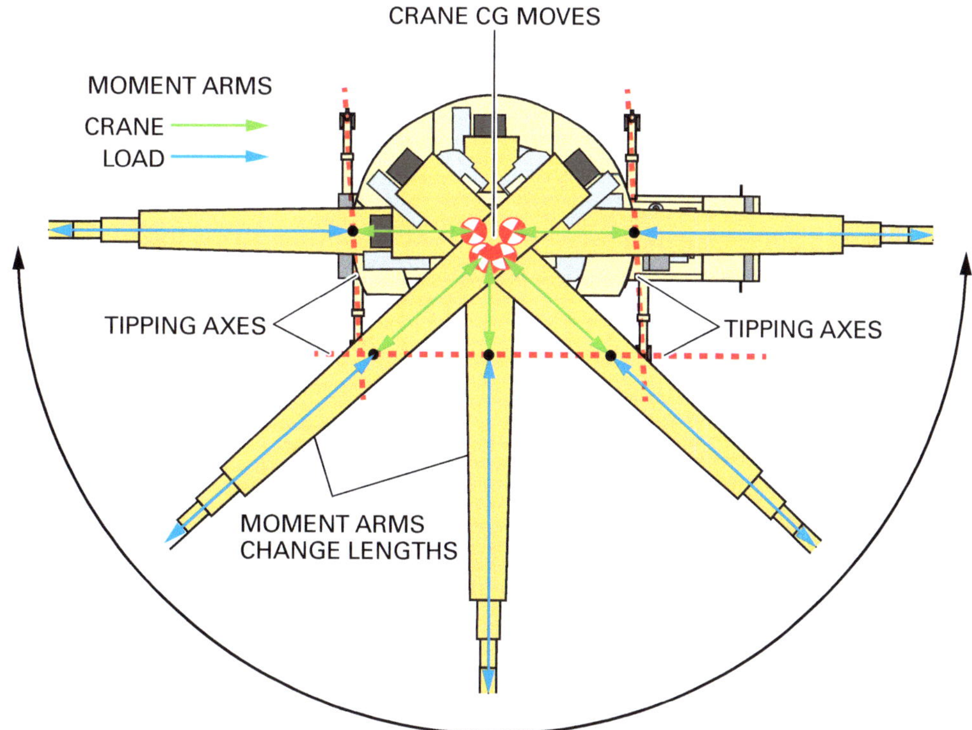

Figure 8 Crane and load moment arms change during swings.

pushes up near the upperworks counterweights or a gust of wind blows. A minor change in any factor could cause the crane to tip.

Therefore, it is crucial that the size of the crane moment always be greater than that of the load moment. This ensures that the crane will not tip. Crane manufacturers develop gross load ratings based on boom length and angle. These yield a load radius that can be used to develop conservative load weight limits that are given on a crane's load chart. This chart takes into consideration the configuration of the crane and the moving centers of gravity while performing a lift. The load chart also accounts for the location of the tipping axes, which will be addressed in the next section.

1.1.4 Acceleration

Before load charts were widely used, operators would estimate crane limitations by watching for signs of tipping. This method of determining capacity limits is dangerous and expensive. Operators using this method probably tipped over more cranes waiting for signs of tipping than they may have saved. The reason for their inability to save a crane after the signs of tipping have occurred is due to a physics principle called acceleration.

Gravity increases the speed of an object in free fall at 32 feet per second (32 ft/s, or 9.8 m/s) for every second it falls. At that acceleration, an object that falls freely straight down for two seconds is traveling 64 ft/s (19.6 m/s) or 44 mph (71 kph).

When a crane starts to tip, the speed at which its CG is moving toward the ground is increasing second by second. The amount of reaction time an operator has to recover from this situation is very limited and requires drastic action with the boom, load, or position of the upperworks. In fact, any indecision or delay on the part of the operator can result in the crane tipping over.

Acceleration can be caused by factors other than gravity. Inattentive operation while hoisting or lowering a load, during swings, or when changing the boom angle or length can all introduce unwanted or excessive acceleration that must be absorbed by the crane. The effects of acceleration and its effects on crane and load moments is discussed in the following sections.

1.2.0 Crane Stability

Crane stability depends on many factors. Even a crane's structural condition can affect its stability. A site that is not level is always a significant factor, as is the condition of the soil beneath the crane. As defined in *ASME Standard B30.5*, a site supervisor exercises control over the work site on which a crane is being used, and over the work being performed on that site. Those responsibilities include being fully aware of any site conditions that may affect a crane lift, such as soft mud or buried utilities. In the absence of a site supervisor—referred to by the term *controlling entity* by OSHA—the employer must assume this responsibility. For taxi cranes, this often means the crane operator must take responsibility for the site conditions as the qualified individual on the site.

There are other factors affecting crane stability that are more related to the physics involved. The factors that will be more closely examined in this section are the quadrants of operation, forward and backward stability, and non-centered lifts.

1.2.1 Quadrants of Operation

As a crane swings to various quadrants, the horizontal distance of the load from the tipping axis in the direction of the boom changes significantly. This is because cranes are typically built to be longer than they are wide to allow them to fit on a roadway. This rectangular-shaped footprint tends to result in more stability of the crane from front-to-rear than from side-to-side.

Crane manufacturers define the quadrants of operations for each of the models that they produce. *Figure 9* shows examples for each type of crane support. In general, wheeled cranes operating on rubber consider any lift to the side of the wheel base as an over-side lift. Load limits for these lifts are typically very restricted because crane moment arms are very short in these directions and tires do not provide a rigid base. On-rubber lifts over the front or rear are the most stable for this type of crane.

The quadrants for wheeled cranes on outriggers are typically defined by diagonal lines passing through the opposite corner floats. The manufacturer may further restrict the size of a quadrant based on how close the crane's CG is to a given tipping axis. Finally, crawler-crane quadrants are defined by diagonal lines passing through the opposite corners where their treads touch the ground. Quadrants of operation also depend on the position of outriggers. Many cranes can have two or more outrigger positions, and the arcs of the quadrants vary depending on the outrigger positions. Note that these quadrants may also be referred to as *defined arcs*.

Along with the definition of the quadrants, the manufacturer also establishes which of the defined areas or arcs are suitable for lifting and specifies any limitations while operating within those quadrants. The quadrants are identified based on the boom's position in relation to the carrier body, and whether the crane is set up on tracks, outriggers, or rubber. The quadrant names are as follows:

- *Over-the-front* – An area located forward of the front outriggers, wheels, or tracks when facing in the forward direction of travel.

- *Over-the-rear* – The area located behind the rear outriggers, wheels, or tracks when facing the rearward direction of travel.
- *Over-the-side* – The area to the left or right side of the crane between the outriggers or beyond the wheels or crawler track.
- *360-degree* – A circular area that encompasses all the other quadrants.

1.2.2 Centers of Gravity and Forward Stability

The movements of the various centers of gravity in relation to the tipping axis while operating a crane affect the crane's forward stability—its resistance to tipping toward the load. The most important thing to be aware of are the positions of the crane's combined CG and the load's CG. As discussed earlier, the crane's CG is a combination of all the centers of gravity of the base section, the upperworks, counterweights, boom, and other components.

Changes that can change the position of the crane's overall CG are:

- Reconfiguring the counterweights
- Extending or retracting the outriggers
- Extending or retracting the crawlers
- Rotating the upperworks
- Extending or retracting the boom
- Raising or lowering the boom

The changes that can affect the distance of the load's CG from the tipping axis are:

- Changing a luffing jib angle.
- Rotating the upperworks
- Extending or retracting the boom
- Raising or lowering the boom

Note that there are several actions that are found on both lists, meaning that they affect both the crane's overall CG as well as the CG of the load.

The relative sizes of the crane and load moments on either side of the tipping axis will determine the stability of the crane. As the crane's CG moves toward the tipping axis, the crane's moment is reduced, often accompanied by an increase of the load's moment. This can also occur if the load is moved further from the tipping axis by extending the boom. Moving the load further from the tipping axis increases the load's moment arm and thus its leverage. Both actions cause the crane to become less stable. The following mathematical examples demonstrate this concept.

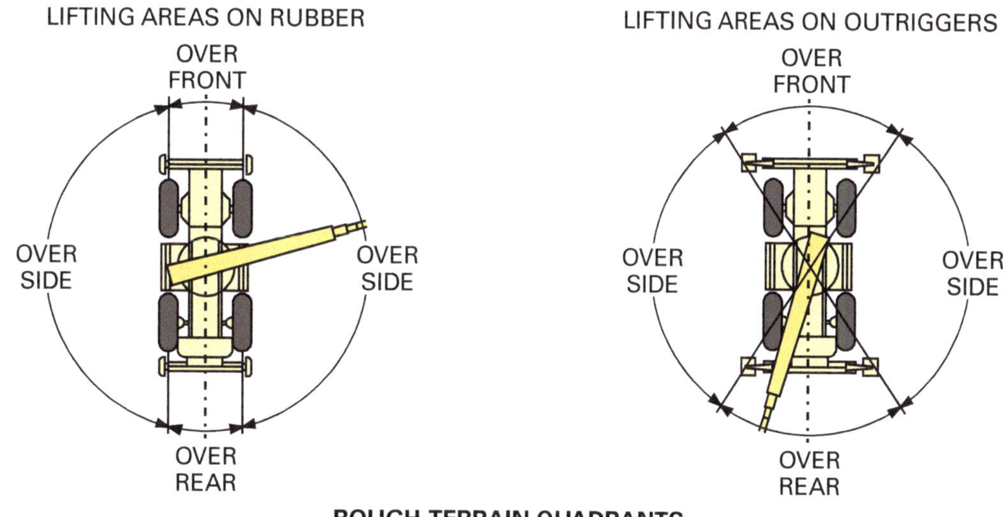

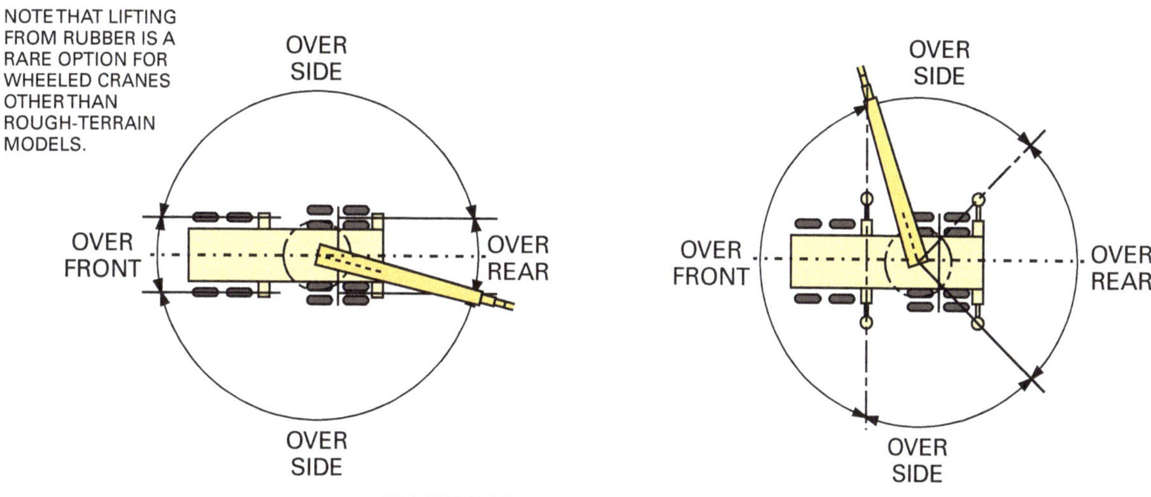

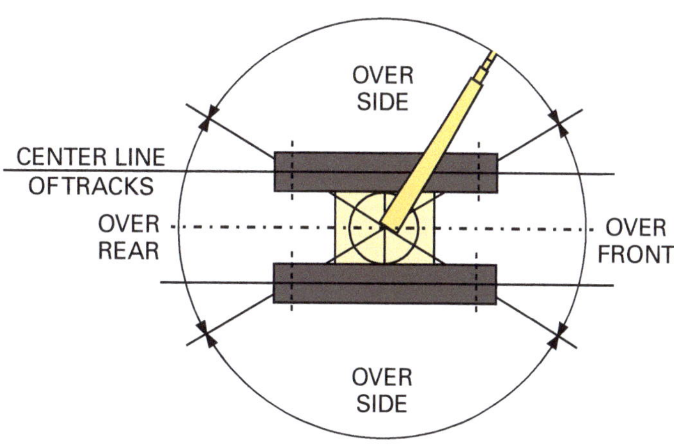

Figure 9 Crane operational quadrants.

Example 1:

Assume a load is being lifted over the front or rear of the crane with the geometry shown in *Figure 10*. The weights and CGs of the crane and load and their distances from the tipping axis are as indicated. The moments applied to each side of the tipping axis are determined by multiplying the weight of the object by its horizontal distance from the tipping axis.

Crane moment:
    4,000 lbs × 7 ft = 28,000 ft·lbs (38,000 N·m)

Load moment:
    500 lbs × 20 ft = 10,000 ft·lbs (13,600 N·m)

The crane moment is much greater than the load moment. The crane is quite stable with 18,000 ft·lb (24,400 N·m) of reserve crane moment, providing a significant margin of safety.

Example 2:

Assume that the boom has now been swung to lift the load from the over-the-side position. The weights of the crane and load remain the same, as shown in *Figure 11*, but the distances of their centers of gravity from the tipping axis have changed. Compute the moments of the crane and load in this position.

Crane moment:
    4,000 lbs × 2.25 ft = 9,000 ft·lbs (12,200 N·m)

Load moment:
    500 lbs × 18 ft = 9,000 ft·lbs (12,200 N·m)

Now the crane's moment is equal to the load moment and there is no margin of safety at all. This lift should not be attempted, as the crane will be very unstable. Note also that the location of the tipping axis relative to the overall crane lever arm moved as the crane was rotated. However, the tipping axis remains between the load and the crane's center of rotation, a fact that will be significant when backward stability is discussed. It is now evident that moving the crane's boom from one operating quadrant to another can have a dramatic effect on crane stability. When an operator takes action that eliminates a margin of safety, it can result in an accident.

1.2.3 Backward Stability

The forward stability of a crane has been discussed in depth. However, the operator must also be concerned with reverse, or backward, stability. The dynamics are only slightly different from those of forward stability. The concern is to keep the total CG of the crane with its load from moving past the tipping axis on the opposite side of

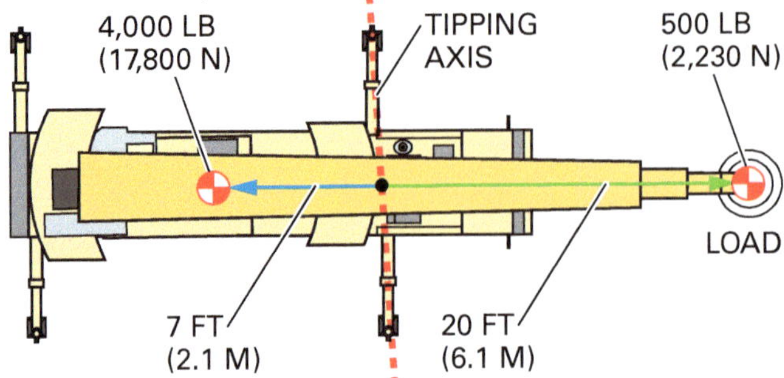

Figure 10 Stability Example 1.

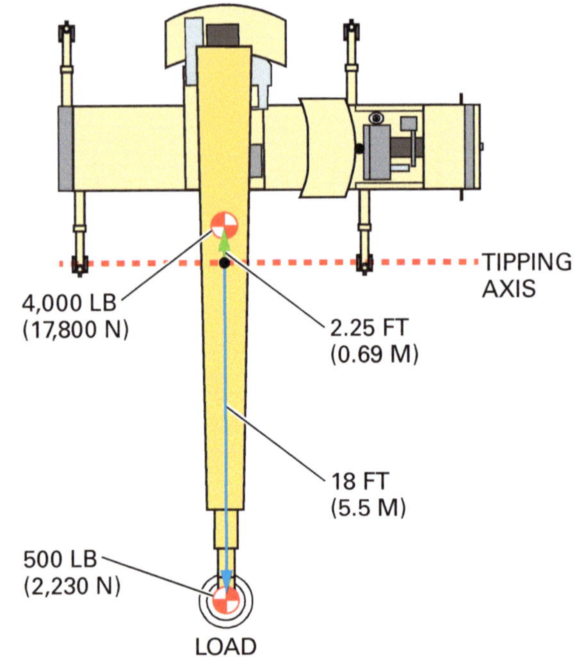

Figure 11 Stability Example 2.

the crane from the load/boom. If this situation occurs, the crane will tip over backwards.

Several factors can contribute to backward tipping. One is the sudden upward movement of the boom. This can occur by either suddenly stopping the boom while raising it or by suddenly releasing the load. In the first case, the boom's momentum is transferred to the crane through its actuators and base. This creates a lever action that tends to tip the crane backward as shown in *Figure 12*. In the other case, the boom springs upward with release of the load's weight, again gaining momentum. The same lever action occurs when the boom suddenly stops against the boom actuators. The upward jerk in the boom, especially at high boom angles, can momentarily shift the crane's combined CG past the point of no return, resulting in the crane tipping backward.

Another factor is the orientation of the crane and boom to unlevel ground. If the operator places a long boom at a high angle on an incline,

NCCER – *Intermediate Rigger*

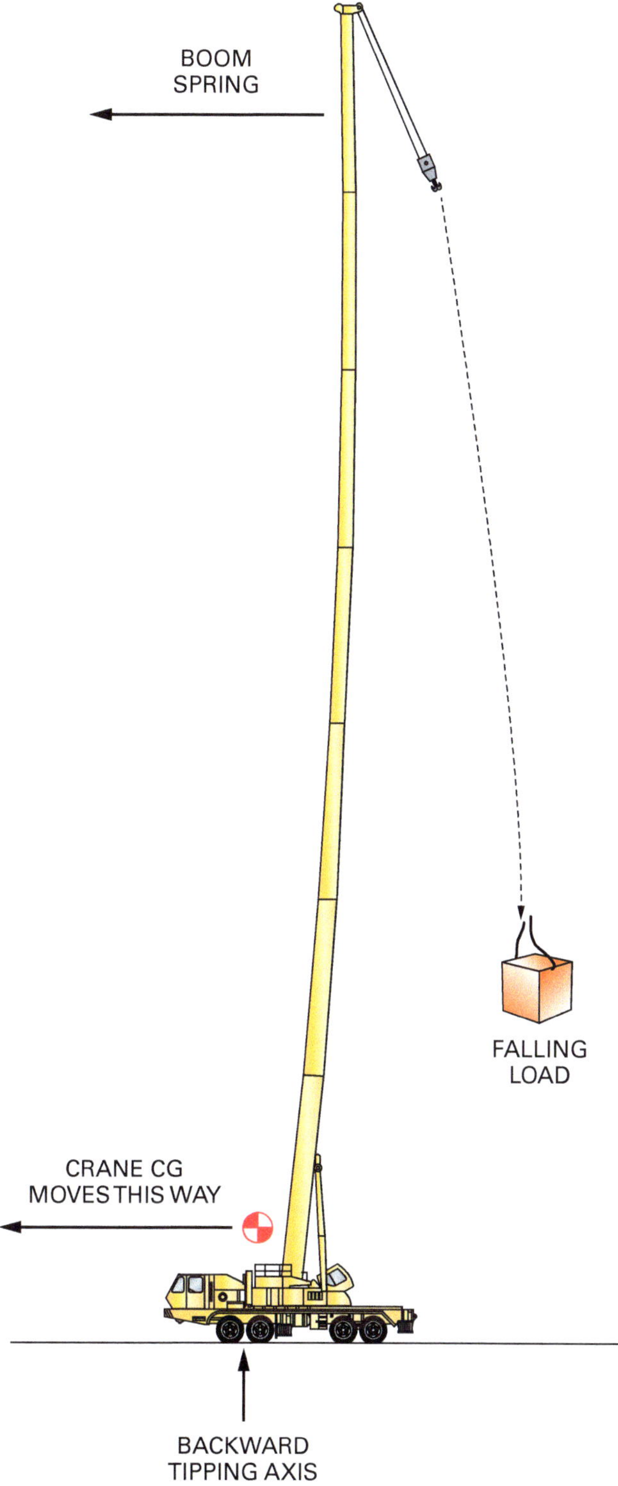

Figure 12 Backward tipping due to boom spring.

these combine to decrease the distance between the crane's center of gravity and rear tipping axis, and reduces the crane's reserve stability. If the crane then travels up the incline, the crane's CG can move past the rear tipping axis, leading to a crane tipping accident.

Rough-terrain (RT) cranes are prone to backward tipping under certain situations. Crane manufacturers are making RT cranes larger and with heavier counterweights. This can create a problem if the operator doesn't follow the manufacturer's notes when swinging the crane boom from over the front to over the side when on rubber. Failure to compensate for heavy counterweights and a short crane moment arm when swinging from one quadrant to another could cause the crane to go over backwards. The same can be said for truck-mounted cranes when operating on rubber and the operator swings over the side. Remember that cranes are typically built to be longer than they are wide, providing more stability from front-to-rear than from side-to-side. A narrow carrier and heavy counterweights can combine to create a backward-tipping risk, especially at high boom angles with no load.

1.2.4 Lifting a Non-Centered Load

A non-centered lift occurs when the load is picked up with the hook and the boom is positioned somewhere other than directly over the load's center of gravity. Such a lift can affect a crane's stability as well as the structural integrity of the boom. When the load is to one side or the other of the boom, side loading of the boom occurs. If the load is farther from the center of rotation than the boom tip, then the load radius will be greater than expected. In either case, as the crane picks up the load, it will result in a larger effective weight of the load, which includes the friction of dragging the load until it is centered under the boom tip.

For loads off to one side of the boom or the other, the load's moment arm is not in line with the boom. As the load is dragged into line with the boom, the change of direction of the moment arm could increase its length relative to the tipping axis. This will decrease the crane's stability until the load and hook blocks have reached a point directly below the boom tip.

Determining a Crane's Backward Stability

ASME Standard B30.5, Mobile and Locomotive Cranes, defines backward stability and how manufacturers test this characteristic. The following text is from *ASME Standard B30.5, Section 5-1.2.2*.

"The backward stability of a crane is its ability to resist overturning in the direction opposite the boom point while in the unloaded condition. The resistance to backward overturning is reflected in the margin of backward stability.

The general conditions for determination of the backward stability margin, applicable to all cranes within the scope of this chapter, are as follows:

(a) crane to be equipped for crane operation with shortest recommended boom

(b) boom positioned at maximum recommended boom angle

(c) crane to be unloaded (lower load block on support)

(d) outriggers free of the bearing surface when the crane is counterweighted for 'on tires or on wheels' operation unless specified by the manufacturer for stationary use

(e) crane to be standing on a firm supporting, level within 1% grade; locomotive cranes to be standing on a level track

(f) all fuel tanks to be at least half full and other fluid levels as specified"

The standard continues with the specific reserve moment contributing to backward stability required for each type of mobile crane. Refer to *ASME Standard B30.5, Section 5-1.2.3* of the ASME standard to review this information.

In some situations, the load may leave the ground before it is under the boom tip. When this happens, the operator must manage the pendulum effect after the load clears the ground, which will cause the load to swing. This creates a very unpredictable and dynamic condition not governed by load charts, resulting in wide variations in boom side loading and/or the actual load radius and moment. These conditions can lead to overloading the crane, causing either a structural failure or a tipping incident.

The pendulum effect is also a factor when swinging the upperworks. Swinging the boom in either direction can cause the load, load block, and hook to swing significantly. Operator trainees will likely practice load control and regaining control of a swinging load during training exercises.

1.3.0 Variables that Influence Stability

Maintaining crane stability is always the primary concern of the crane operator, and the variables addressed in the previous section must always be considered prior to and during every lift. Some less common but equally important variables are discussed in this section.

1.3.1 Wind

One of the most overlooked stability factors affecting cranes is the wind (*Figure 13*). High wind speeds can dramatically affect a crane and its load. Almost all crane manufacturers specify in the load chart that chart ratings must be reduced under windy conditions, and they may also recommend a shutdown wind speed. In many cases, when the wind speed exceeds 30 mph (48 kph), it is advisable to stop operations.

Wind impacts both the crane and the load, affecting the crane's load moment and thus the rated capacity of the crane. Use a great deal of discretion when lifting with even moderate wind speeds that a crane's load chart allows, especially if the winds are gusty. It is imperative the crane operator know the precise out-of-service wind speed for the crane as specified by the manufacturer. When this information is ignored, there can be consequences that include equipment damage, injuries, and/or loss of life. Consult the wind speed charts as necessary and follow them without fail.

An example of a wind speed chart for a given crane is shown in *Table 1*. Note that some cities and counties have specific regulations regarding wind speed that may exceed the manufacturer's requirements. If a jobsite or locality imposes a wind speed limit that differs from the manufacturer's charts, the lowest of the two values is the limit.

> **CAUTION**
>
> The wind chart shown in *Table 1* is for a specific crane model and cannot be applied to all cranes. Always check and follow the manufacturer's wind chart speed for the specific crane in use.

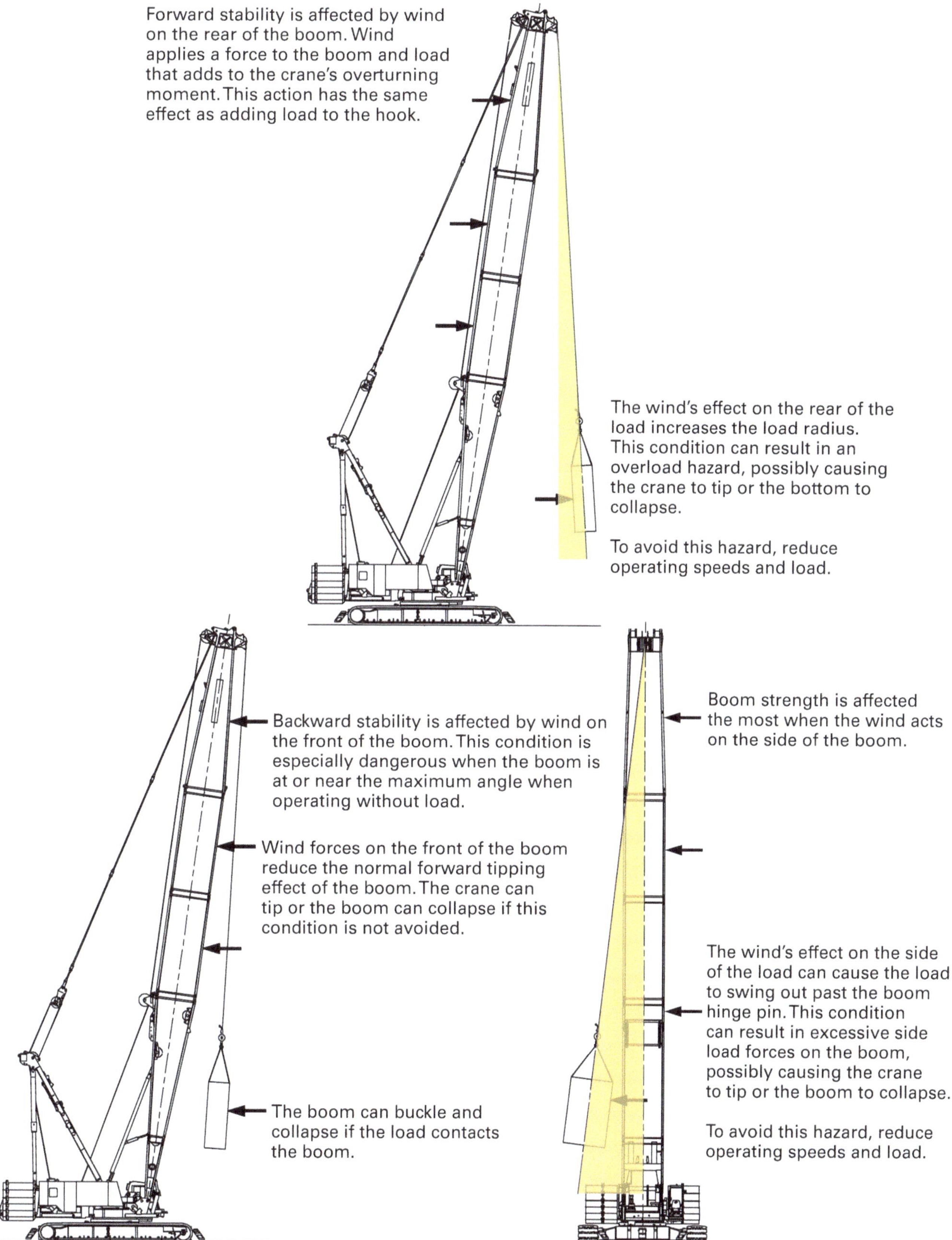

Figure 13 Wind and its effect on crane stability.

Table 1 Example of a Wind Speed Chart

Boom and Boom + Jib Lengths up to 250'	
Description	**Allowable Windspeeds in Miles Per Hour (mph)**
1. Normal Lifting Operation. (See Capacity Charts.)	0–20 mph
2. Reduced Operation. Capacities must be reduced by 20%.	21–30 mph
3. Reduced Operation. Capacities must be reduced by 40%.	31–40 mph
4. Reduced Operation. Capacities must be reduced by 70%.	41–45 mph
5. No Operation. Store attachment on ground.	Over 45 mph
Boom and Boom + Jib Lengths Greater than to 250'	
1. Normal Lifting Operation. (See Capacity Charts.)	0–20 mph
2. Reduced Operation. Capacities must be reduced by 35%.	21–30 mph
3. Reduced Operation. Capacities must be reduced by 60%.	31–40 mph
4. Reduced Operation. Capacities must be reduced by 70%.	41–45 mph
5. No Operation. Store attachment on ground.	Over 45 mph

The wind-catching surface of the load must be considered whenever wind is present as control of the load can be lost. A 20-mph (32-kph) wind exerts a pressure of only 1.2 lb/ft² (58 N/m²) on a flat-surfaced load. The force exerted on a common sheet of plywood is 38 lb (167 N), for example. Therefore, under these conditions, only loads having a large sail area—area that would catch the wind—would require the crane's capacity to be derated. At 30 mph (48 kph), however, the wind exerts a pressure of 2.6 lb/ft² (125 N/m²) on the same flat surface area, more than double that of a 20-mph wind. This results in an 84-lb (375-N) force exerted on a common sheet of plywood. This much wind is enough to cause load control problems.

Forward stability is the critical consideration when the wind is coming from behind the boom. It applies a force to the boom and to the load that adds to the tipping moment of the crane. This has the same effect as adding load to the hook. Backward stability is the critical factor when the wind is from the front, particularly when the boom is at or approaching the maximum boom angle. This has the same effect as reducing the load on the hook. The wind forces on the boom reduce the forward moment normally provided by the boom by lifting it higher and closer to the tipping axis.

1.3.2 Impact Forces

Another equally dangerous dynamic effect is known as impact loading or dynamic loading. An impact load is the force exerted by something on a moving object to stop it or slow it down as they collide. A large, short-duration force is required to eliminate the momentum of the moving object by stopping it. Impact loading occurs, for example, when a free-falling load is suddenly snubbed by the hoist drum brake. Snatching a load off the ground will also cause an impact load to occur.

Impact loading causes an instantaneous increase in effective load weight. When a load is allowed to drop freely, the boom will initially recoil upward as the drop begins, but when the hoist drum is braked or locked to stop the descent, the full weight of the load plus the impact loading to stop its movement is suddenly applied to the crane through the wire rope. The result can be the overturning of the crane or structural failure of the boom.

NCCER – *Intermediate Rigger*

Avoiding the Jerks in Your Life

With a *Crack!*, the batter sends the fastball into the outfield. What's happening during that instant when the bat strikes the baseball? The pitched ball, before its impact, has a velocity that includes both speed and direction. As the bat hits the ball, it exerts a very large force to change its direction and give it a different speed.

All moving objects have mass (m) and velocity (v). The product of these equals the momentum (M) of the object: $M = m \times v$. A force is required to change an object's momentum, which is normally done by just changing its velocity, since an object's mass usually remains unchanged. Velocity change can be either a change in speed, a change in direction, or both, as in the case of the struck baseball.

A rapid change in momentum, as occurs with a bat hitting a ball, is referred to as an *impulse* or an *impact*, and the force required to make the change is an impulsive or impact force. Mathematically, the change in momentum (ΔM) can be described by the equation,

$$\Delta M = F\Delta t$$

In this equation, ΔM is the change in momentum, F is the force, and Δt is the time interval during which the force is exerted. Notice that, for a given change in momentum, the shorter the time elapsed, the greater the force required to make it happen. For this reason, sharp jerks or snatching of the load must be avoided. Stretching out changes in the load's momentum over time (moving slowly) greatly reduces the changes in effective load weights.

Similar changes to effective load weights (and the load moments) that result from impact loading occur during a significant acceleration or deceleration of a load. A rapid hoisting acceleration will produce a hook load greater than the actual load weight. If there is a sudden release of a lifted load, the action can cause the boom to recoil as described previously. If this occurs at a high boom angle, the crane can topple over backwards. The crane operator should remember to gradually transfer the load's weight to the crane when hoisting. When lowering, the load should be placed gently to allow boom deflection and pendant stretch to gradually return to normal.

1.3.3 Submerged Lifts

Submerged lifts involve a load that is completely or partially immersed in a liquid during the operation. Lifting submerged or immersed loads can be hazardous if all environmental factors affecting the load are not considered. Even before a lift can begin, properly siting the crane must be the first consideration.

Assuming the crane is to operate from land, ground stability near the water's edge must be evaluated. Ground along unimproved shorelines can be saturated and very unstable. Mats or blocking may be required as part of the lift preparations to ensure adequate ground support near water. Docks and piers must be certified by competent individuals before even driving a crane onto them. Conducting a lift from these structures requires a careful analysis of the structure.

Mobile cranes may be placed on barges for submerged work (*Figure 14*). The extra requirements and safety precautions required for floating crane lifts are covered in *ASME Standard B30.8, Floating Cranes and Floating Derricks* and 29 *CFR* 1926.1437, *Floating Cranes/Derricks and Land Cranes/Derricks on Barges*. The OSHA standard does not apply to cranes on jacked barges fully supported underneath, which are not technically floating.

The main factor to be aware of during submerged lifts is how buoyancy can affect the effective weight of the load at various stages of the lift. Buoyancy is the upward force that a liquid or gas exerts on an object immersed in the fluid. For an object submerged in water, the buoyant force is equal to the weight of water displaced by the wetted volume of the object. The relationship of the weight of an object to the weight of water it displaces determines the significance of this effect. If the weight of water displaced by a completely submerged object is greater than its weight, then the object (like a boat or ship) will rise at the surface of the water until its immersed displacement exactly equals its weight. In this case, the object is only immersed in the water, not submerged. Its effective weight while floating is zero—the difference of its actual weight and the water it displaces.

If the weight of water displaced by a completely submerged object is less than its weight, then the object—a rock, for example—will sink to the bottom. However, its effective weight while submerged is still the difference of its actual weight and the buoyant force of the water it displaces.

Figure 14 Mobile crane on a barge conducting submerged lifts.

If the weight of the completely submerged object is exactly equal to the weight of the displaced water, it will neither float nor sink—it is neutrally buoyant. In this condition, the submerged object has zero effective weight. However, it still has mass and inertia, so trying to move a neutrally-buoyant object can take a substantial force.

The most critical point in a submerged lift is when the load is brought to the surface of the water. When the load emerges, the volume and weight of water it displaces begins to decrease. As the buoyant force drops, the load's effective weight increases. When the load lifts free from the water, the full weight of the load will be transferred to the crane. If this change in weight is not anticipated in the lift calculations, the crane can be significantly overloaded, resulting in structural failure or tipping.

Consequently, submerged-lift planning must account for the full weight of the load in air with plenty of margin included. Note that a submerged porous load, such as waterlogged wood, or a hollow object full of water, will be far heavier than their theoretical dry weights. After lifting free from the water's surface, trapped water may drain out partially or completely, dynamically changing the load's weight during the lift.

When placing an object into the water, the opposite effects are observed. An object such as the artificial reef being placed in *Figure 15* may initially float, then fill with water, becoming much heavier. The dynamics of the lifting operation must be continually monitored by the operator to avoid an accident.

Another condition to consider is wave action. As the load comes to the surface during a lift, the rise and fall of the water's surface will change the volume of the object submerged as every second passes. The changes in buoyancy will impose load accelerations on the crane. These changes can cause instability of the crane or overload its various components.

A side effect of submerged lifts, as well as lifts from wet areas, is the suction effect. An object resting in a mucky bottom or on wet ground forms a wet seal with its surroundings. To lift the object, a void space must be formed underneath it so it can rise free. If the void is easily filled with water or air, then the object lifts easily. However, if the wet seal prevents water, air, or even the soil itself from sliding past the object to fill the void, a pressure difference is created—suction—and the object is stuck. The suction resistance can add unexpected weight to the crane. With enough lifting force, the suction will eventually break free, suddenly reducing the effective weight of the load. This unexpected loss of load deflects the boom backwards, possibly resulting in the crane tipping backward or structural stress on the boom as described earlier.

Submerged lifts should always be well-planned and carefully executed. At times, the lift is best

Figure 15 Crane placing a discarded aircraft as an artificial reef.

performed in two steps. The first step is to lift one end of the load to break the suction, using leverage to increase the crane's lifting force on the object. The next step involves lifting the load once it is free from the suction. Newer cranes have a sophisticated **load moment indicator (LMI)** that allows the operator to know how much lifting force is applied to the load. This is a much safer way of breaking a suction than to pull until the suction is broken. Using an LMI should minimize the risk of overloading the crane, which could result in a crane accident. The use of older cranes without an LMI should be discouraged for submerged lifts because the operator will need to rely on the line pull of the crane with no idea of how much force is really being exerted.

Additional Resources

ASME Standard B30.5, Mobile and Locomotive Cranes. Current edition. New York, NY: American Society of Mechanical Engineers.

ASME Standard B30.8, Floating Cranes and Floating Derricks. Current edition. New York, NY: American Society of Mechanical Engineers.

29 *CFR* 1926, Subpart CC,*Cranes and Derricks in Construction.* Current edition. Washington, DC: US Department of Labor, Occupational Safety and Health Administration.

Mobile Crane Safety Manual. 2014. Milwaukee, WI: Association of Equipment Manufacturers.

Cranes: Design, Practice, and Maintenance. Second edition; 2002. Ing J. Verschoof. Hoboken, NJ: John Wiley and Sons, Inc.

IPT's Crane and Rigging Handbook. Current edition. Ronald G. Garby. Spruce Grove, Alberta, Canada: IPT Publishing and Training Ltd.

North American Crane Bureau, Inc. website offers resources for products and training, **www.cranesafe.com**.

1.0.0 Section Review

1. Which of the following factors plays a part in the function of a lever?

   a. How long the force is applied
   b. The length of the lever arm
   c. The weight of the fulcrum
   d. The size of the fulcrum

2. Which of the following is *not* a factor in determining the different quadrants of operation for a particular crane?

   a. The type of crane
   b. The load's CG
   c. The crane's points of contact with the ground
   d. The use of outriggers

3. Which of the following statements regarding dynamic loads during a lift is *correct*?

   a. Dropping a lifted load poses no hazard to the crane.
   b. Load charts include a margin for catching a falling load with the hoist drum brake.
   c. Dropping a lifted load causes the boom to initially recoil upward.
   d. The force required to stop a falling load cannot exceed the full weight of the load.

SUMMARY

The factors that influence a lifting operation are numerous, and a change in any one of these factors can cause the crane and the load to react. It is imperative that crane operators understand and consider the different load dynamics. Operators must know how to prevent tipping conditions and, if necessary, how to react in the event such conditions occur. They must be aware of how the crane and load moments can change simply by rotating the crane through its operating quadrants. Operators working on or around water conducting submerged lifts must also be aware of the effects of buoyancy and wave action, in addition to wind and other factors common to all lifts.

Today's cranes are extremely sophisticated and safer than ever, but they require alert and aware operators. A properly trained operator will ensure that the stability and structural integrity of the crane will not be compromised. This, in turn, will help to ensure the safety of the load and the lift team.

1. In crane operations, the lever principle mainly applies to _____.

   a. raising and lowering the lifting hoist
   b. crane stability when lifting
   c. operation of the crawler tracks
   d. power flow from the engine

2. The unit used to express a force's moment is _____.

   a. lb/ft^2
   b. kPa/m^2
   c. ft·lb
   d. kg·m/s

3. Which of the following statements is *correct*?

   a. The CG is the point where an object's mass is assumed to be concentrated.
   b. The crane stability model combines the crane CG and load CG into a single CG.
   c. To be stable, the crane CG must always be on the load side of the tipping axis.
   d. For stability, the crane moment must equal the load moment.

4. The type of crane that has the most limitations within its quadrants of operation is the _____.

   a. wheeled crane operating on rubber
   b. rough-terrain cranes on outriggers
   c. all-terrain crane on outriggers
   d. crawler crane on extended tracks

5. As a crane's CG moves in the direction of the tipping axis, the _____.

   a. crane's moment is increased
   b. crane's moment is reduced
   c. crane is less likely to tip towards the load
   d. crane is more likely to tip backwards

6. Refer to the crane stability model shown in *Figure RQ01*. Determine which statement below accurately describes the stability of the crane in this configuration.

   a. The crane is unstable and will tip toward the load.
   b. The crane is stable but with no reserve moment.
   c. The crane is stable with a large reserve moment.
   d. The crane is unstable in the backward direction.

7. If a jobsite or locality imposes a wind speed limit that differs from the crane manufacturer's charts, the _____.

   a. highest wind speed of the two values is the limit
   b. lowest wind speed of the two values is the limit
   c. local limit is the effective limit, regardless of the manufacturer's charts
   d. manufacturer's limit is the effective limit, regardless of local or jobsite regulations

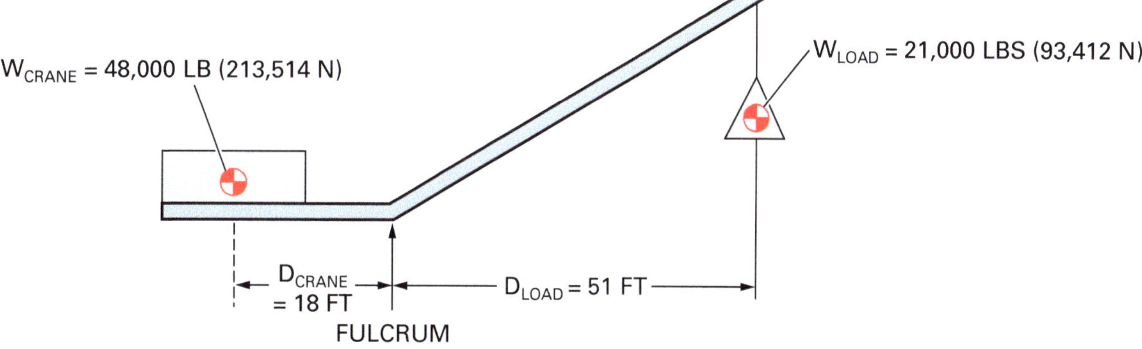

Figure RQ01

8. Which situation could potentially lead to a crane tipping over backward?

    a. Wind from behind and nominal boom angle with the hoist attached to the load on the ground
    b. Wind from the front and the boom at its maximum angle
    c. Wind from the side with no suspended load
    d. Wind from behind with a nominal boom angle and no suspended load

9. Winds of 30 mph (48 kph) exert a pressure on a flat surface that is _____.

    a. more than twice the force applied by a 20-mph (32-kph) wind
    b. about 1.2 lb/ft^2(58 N/m^2)
    c. acceptable if the load weight on the hook is increased by 40 percent
    d. only dangerous if the boom and jib length is greater than 250 feet (76 meters)

10. For lifts involving submerged objects near unimproved shorelines, what is the first thing an operator must evaluate?

    a. The depth of the water
    b. The clarity of the water
    c. The type of bottom underwater
    d. The stability of the ground from which the crane will operate

Trade Terms Introduced in This Module

Acceleration: In common usage, the rate of increase of an object's speed per second. The rate of decrease of speed is commonly called *deceleration*.

Buoyancy: The upward force exerted by a liquid or gas (both fluids) on an object immersed in the fluid.

Center of gravity (CG): The point at which the entire weight of an object is concentrated, such that supporting the object above this specific point would result in its remaining balanced in position.

Center of rotation: In crane operations, the vertical axis through the center of the swing circle around which the upperworks rotates.

Crane moment: The product of the entire crane's center of gravity and its distance from the tipping axis along the lever arm toward the load's center of gravity. Commonly called the crane's *leverage* or *lever force*.

Dynamics: A general term that refers to forces, their effects, and how they change from one moment to the next.

Effective weight: Also referred to as the apparent weight. The weight of a load plus or minus the effect of other factors, such as acceleration, wind, and buoyancy. For example, a load that accelerates while falling exerts a force that is greater than the actual weight of the load when it is suddenly stopped.

Fulcrum: The point of support on which a lever pivots.

Impact loading: A short-duration force that is suddenly exerted when a moving object is slowed or stopped.

Law of moments: A scientific principle that defines the conditions under which a lever-and-fulcrum system is balanced and stable.

Lever: A simple machine consisting of a rigid arm pivoting on a fulcrum. Force can be transferred from one end to the other as it pivots, changing the force's direction and size, or two forces acting on its ends can work against each other.

Leverage: A common term describing the action of a lever, especially for prying or lifting a heavy object. See *moment*.

Load moment: The force applied to the crane by the load; the leverage of the load, opposing the leverage of the crane. The load moment is calculated by multiplying the gross load weight by the horizontal distance from the tipping fulcrum to the center of gravity of the suspended load. The load moment is usually reported to the operator as a percentage of the crane's capacity at the present set of conditions. As those conditions change, such as the boom angle, the load moment changes as well.

Load moment indicator (LMI): A system that aids the equipment operator by directly or indirectly sensing the overturning moment on the equipment. It compares the lifting conditions to the equipment's rated capacity, and when the rated capacity is reached, it disables equipment functions that can increase the severity of loading on the equipment, such as hoisting or telescoping the boom further out.

Load radius: The horizontal distance between the crane's center of rotation and the vertical hoist line or tackle with the load applied. On load charts, the load radius is determined by the boom angle and its length to the boom tip.

Moment: The effect of a force acting perpendicularly on a lever at a certain distance from its fulcrum. Commonly called *leverage*. Mathematically, moment is equal to the product of the force and its moment arm.

Moment arm: In a lever system, the distance between the point where a force is applied to a lever and the lever's fulcrum.

Momentum: A property of all moving objects that is equal to the product of their mass and velocity. In equation form, $M = m \times v = mv$, where M is the object's momentum, m is its mass, and v is its velocity.

Pendulum effect: The variable force exerted by a swinging load on the crane structure along the hoist line, like the swing of a pendulum. The load's excess force is maximum at the bottom of the swing.

Stability: A measure of a movable object's ability to resist a change in position of its center of gravity. The more force required to raise its CG, the more stable it is.

Tipping axis: In crane operations, an imaginary horizontal line around which a crane's center of gravity would rotate as it tips under the right circumstances.

Additional Resources

This module is intended as a thorough resource for task training. The following reference material is recommended for further study.

ASME Standard B30.5, Mobile and Locomotive Cranes. Current edition. New York, NY: American Society of Mechanical Engineers.

ASME Standard B30.8, Floating Cranes and Floating Derricks. Current edition. New York, NY: American Society of Mechanical Engineers.

29 *CFR* 1926.1400, *Cranes and Derricks in Construction*. Current edition. Washington, DC: US Department of Labor, Occupational Safety and Health Administration.

Mobile Crane Safety Manual. 2014. Milwaukee, WI: Association of Equipment Manufacturers.

Cranes: Design, Practice, and Maintenance. Second edition; 2002. Ing J. Verschoof. Hoboken, NJ: John Wiley and Sons, Inc.

IPT's Crane and Rigging Handbook. Current Edition. Ronald G. Garby. Spruce Grove, Alberta, Canada: IPT Publishing and Training Ltd.

North American Crane Bureau, Inc. website offers resources for products and training, **www.cranesafe.com**.

Figure Credits

Section Review Answer Key

Answer Section One	Section Reference	Objective
1. b	1.1.1	1a
2. b	1.2.1	1b
3. c	1.3.2	1c

NCCER CURRICULA — USER UPDATE

NCCER makes every effort to keep its textbooks up-to-date and free of technical errors. We appreciate your help in this process. If you find an error, a typographical mistake, or an inaccuracy in NCCER's curricula, please fill out this form (or a photocopy), or complete the online form at **www.nccer.org/olf**. Be sure to include the exact module ID number, page number, a detailed description, and your recommended correction. Your input will be brought to the attention of the Authoring Team. Thank you for your assistance.

Instructors – If you have an idea for improving this textbook, or have found that additional materials were necessary to teach this module effectively, please let us know so that we may present your suggestions to the Authoring Team.

NCCER Product Development and Revision

13614 Progress Blvd., Alachua, FL 32615

Email: curriculum@nccer.org
Online: www.nccer.org/olf

❏ Trainee Guide   ❏ Lesson Plans   ❏ Exam   ❏ PowerPoints   Other ___________________

Craft / Level: ____ Copyright Date: _______________

Module ID Number / Title: ____

Section Number(s): ____

Description: ____

____

____

____

Recommended Correction: ____

____

____

____

Your Name: ____

Address: ____

Email: ____ Phone: _________________________

Wire Rope

OVERVIEW

Wire rope serves as the crucial link between the crane and its load. The well-known saying, "a chain is only as strong as its weakest link," certainly applies to wire rope as well. This crane component, appropriately considered to be a machine of its own, must be properly applied and cared for throughout its service life. This module presents the construction features of wire rope products and provides guidance for its inspection, maintenance, and handling.

Module 21204

Trainees with successful module completions may be eligible for credentialing through the NCCER Registry. To learn more, go to **www.nccer.org** or contact us at 1.888.622.3720. Our website has information on the latest product releases and training, as well as online versions of our *Cornerstone* magazine and Pearson's product catalog.

Your feedback is welcome. You may email your comments to **curriculum@nccer.org**, send general comments and inquiries to **info@nccer.org**, or fill in the User Update form at the back of this module.

This information is general in nature and intended for training purposes only. Actual performance of activities described in this manual requires compliance with all applicable operating, service, maintenance, and safety procedures under the direction of qualified personnel. References in this manual to patented or proprietary devices do not constitute a recommendation of their use.

Objectives

When you have completed this module, you will be able to do the following:

1. Describe the construction and configurations of wire rope and explain how to create terminations.
   a. Describe the construction and configurations of wire rope.
   b. Explain how to create various wire rope terminations.
2. Identify and describe wire rope inspection, handling, and maintenance requirements.
   a. Identify standards-based wire rope inspection requirements and the actions to be taken for various deficiencies.
   b. Describe the various types of wire rope deficiencies that may be encountered.
   c. Describe how to inspect sheaves and load blocks.
   d. Describe common wire rope handling and maintenance procedures.
3. Explain how to reeve wire rope onto load blocks and drums.
   a. Identify important reeving considerations.
   b. Explain how to reeve wire rope onto a hoist drum.

Performance Tasks

Under the supervision of your instructor, you should be able to do the following:

1. Assemble a conventional wedge socket.
2. Inspect wire rope using the appropriate inspection criteria.
3. Inspect a sheave and load block using the appropriate inspection criteria.
4. Reeve multiple-part wire rope to a load block.

Trade Terms

Birdcaging
Constructional stretch
Core
Crossover points
D/d ratio
Dead end
Design factor
Fatigue fractures
Fiber core (FC)

Filler wire
Flange points
Independent wire rope
   core (IWRC)
Lay
Lay length
Live section
Minimum breaking
   strength

Monel™
Pendants
Pitch diameter
Preformed strand
Preformed wire rope
Repetitive pickup
   points
Rotation-resistant wire
   rope

Running ropes
Seizing
Speltered
Strands
Strand patterns
Swaged
Wire
Wire strand core (WSC)

Industry Recognized Credentials

Contents

1.0.0 Wire Rope Construction and Terminations

Objective

Describe the construction and configurations of wire rope and explain how to create terminations.

a. Describe the construction and configurations of wire rope.
b. Explain how to create various wire rope terminations.

Performance Tasks

1. Assemble a conventional wedge socket.

Trade Terms

Birdcaging: A term used to describe the condition where wire rope is forced into compression along its length, pushing the outer strands away from the core and/or inner strands, forming a shape that resembles a bird cage.

Core: The axial part at the center of wire rope around which the individual strands are laid.

Dead end: The excess wire rope exposed after a termination has been properly fitted. This part of the wire rope is not exposed to the strain applied by a load to the wire rope.

Fiber core (FC): A cord or rope of natural or synthetic fiber used as the axial center (core) of a wire rope.

Filler wire: Small wire used to create spacers within a strand that help hold the position and support the other strand wires.

Independent wire rope core (IWRC): A wire rope used as the axial center (core) of a larger wire rope assembly.

Lay: In the context of wire rope, the lay describes the direction of rotation of the strands in a wire rope assembly.

Lay length: The distance measured parallel to the axis of the rope or strand in which a strand or wire makes one complete revolution around the core or center.

Live section: Any part of a wire rope that is exposed to the load weight.

Minimum breaking strength: Also referred to as *minimum breaking force*. The documented breaking strength of a wire rope product established under set industry testing procedures.

Monel™: A trademarked metal alloy made primarily from nickel and copper, with small amounts of other materials added. Monel is resistant to corrosion initiated by a number of common sources, including seawater, and is an expensive material.

Preformed strand: A strand constructed of wires that were formed into the helical shape they would assume once placed in the strand, before the strand was assembled.

Preformed wire rope: A wire rope constructed of strands that were formed into the helical shape they would assume once placed in the rope, before the rope was assembled.

Rotation-resistant wire rope: Wire rope constructed with inner and outer strands placed in opposing lays, rather than using the same lay in both layers. This causes the torque of the strand layers to oppose each other, significantly reducing the tendency to twist.

Seizing: The securing of an open end of wire rope by wrapping wire tightly around its circumference several times. This action prevents fraying and unraveling of the wire rope. This preparation is used on both sides of a planned wire rope cut.

Speltered: To be attached with molten zinc or a zinc alloy. The term spelter is considered a synonym for zinc.

Strands: A single grouping of round or shaped wires laid helically around a core or center wire. A wire rope assembly for lifting purposes is typically a collection of strands around a common core.

Strand patterns: The various configurations of the wires used to form a strand. The four basic strand patterns are wire, Seale, Warrington, and filler wire.

Swaged: A connection made by forming metal, usually with a die or shaped tool, to change the diameter. A swaged wire rope connection is made by reducing the diameter of the socket around the rope.

Wire: A single, continuous length of metal, usually cold-drawn or cold-rolled from a rod, available in various diameters.

Wire strand core (WSC): A wire strand assembly used as the axial center (core) member of a wire rope.

A mobile crane has enormous power that can be used to lift loads and move them to a desired location. To lift a load, the crane must be able to transfer this power into a lifting force that acts on the load. A crane transfers rotational power to the hoist drum, which in turn transfers lifting power to the load through a wire rope. Wire rope is a very sophisticated component that is considered a machine in its own right. Many types of wire rope are specifically designed to support various mobile crane operation requirements. This module presents information about the construction, care, inspection, and use of wire rope.

Replacement wire rope must have a strength rating at least as great as the original rope furnished, or that recommended by the manufacturer. Any change from the original size, grade, or construction is to be approved by a rope manufacturer, the crane manufacturer, or an otherwise qualified person. Using a wire rope for hoisting that does not have the proper construction for the application can lead to unexpected rope failures, significant property damage, and serious personal injury.

1.1.0 Wire Rope Construction

An understanding of the construction of wire rope and its components is critical for effective inspections and proper care. There are hundreds of different kinds of wire rope on the market. All types of wire rope are not created to do precisely the same job under the same conditions. This section provides information on wire rope construction and explains how the types differ. Since the performance of each type is different, wire rope must be selected to meet the needs of the crane and its design.

As a general rule, a product must be greater than $^3/_8$" in diameter to be considered wire rope. Smaller sizes are generally referred to as *cable*. A wire rope consists of only three main parts: the wire, the strands, and the core, as shown in *Figure 1*. The wire represents the smallest individual component of a wire rope assembly. The method of construction and working relationship between the parts gives each specific type of rope its unique characteristics. Each of the components that comprise wire rope interact with each other in some way as the rope bends or is stressed. Interactions occur within the individual strands as well as between each of the strands that make up the product. It is this consistent interaction that allows wire rope to be classified as a machine.

The wire can be made of many different materials and in a variety of shapes. It can be made of metals such as steel, iron, bronze, and Monel™. Because of the wide variety and availability of various grades of steel, it is the most common type of material used in the construction of wire rope. The most common grades of steel used for wire rope, in order of increasing breaking strength, are as follows:

- *Traction steel (TS)* – Used in elevator service.
- *Plow steel (PS)* – Has low tensile strength, but can be used when strength is secondary to wear resistance.
- *Improved plow steel (IPS)* – The most commonly used for mobile crane wire rope. It is wear-resistant, has high tensile strength, and is fatigue-resistant.
- *Extra improved plow steel (EIPS)* – Used when the need for higher breaking strength is required. The bending quality of this rope is not as good as that of improved plow steel.
- *Extra extra improved plow steel (EEIPS)* – Used when special installations require maximum breaking strength, such as mine-shaft hoisting.

Note that there may be other acronyms used to identify these types of steel, and to identify wire ropes constructed of other metals. For example, extra improved plow steel may also be referred to as XIPS or XIP, instead of EIPS.

The term *plow steel* is a reference to an old grading system for steel developed to provide a consistent gauge to determine the tensile strengths of steel. This term is still used today as a reference for grading wire rope.

Another descriptive characteristic of wire rope is the finish of the wire. Most wire is made from high-carbon steel. Unless a special coating is added to change the finish, it will have a bright

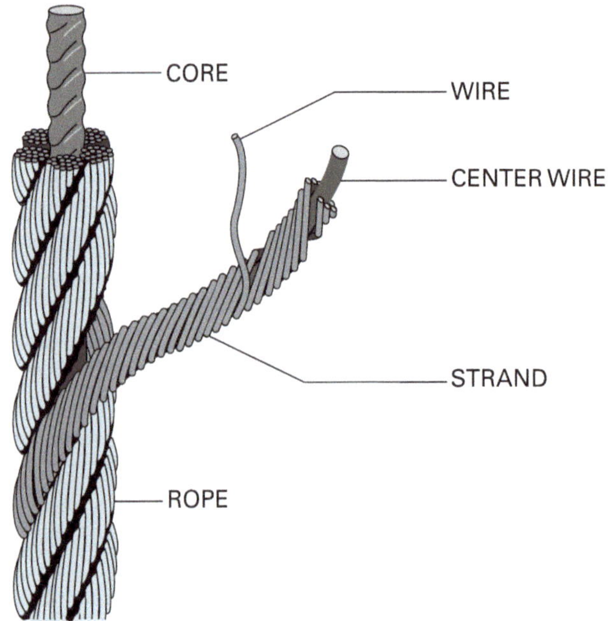

Figure 1 Components of a wire rope.

finish. Some wire ropes are manufactured with special protective coatings such as zinc, resin, or even plastic.

Each strand of a wire rope is made up of two or more wires assembled in a specific pattern. Strands can be assembled from several types of material in addition to steel wire. The materials of construction may include natural or synthetic fibers, even in a wire rope product. Wire ropes can be made from any number of strands fitted together.

The core provides a foundation and support for the collection of strands. The core can be made of several different materials, including steel and synthetic fibers. Typical wire rope core types are identified as wire strand core (WSC), independent wire rope core (IWRC), and fiber core (FC). A wire strand core is exactly as it sounds—a strand that serves as the rope's core. The independent wire rope core is also exactly as it sounds—a core that is a complete wire rope assembly consisting of strands and its own core. IWRC is the most common type used for hoisting. Fiber core wire ropes are those products that use natural or synthetic fibers to form the core.

Synthetic Rope

Synthetic rope has been developed for use in crane applications, and the product is readily available. However, it has not yet received widespread acceptance in the industry. The high-performance synthetic fibers used to manufacture synthetic rope provide lifting capacity that is similar to that of wire rope of the same size. In the photo shown here, the outer portion of the rope has been compressed to expose the core.

There are applications where synthetic rope excels. Submerged lifts, for example, can accelerate the corrosion and degradation of wire rope, especially in salt water. Synthetic rope is not subject to corrosion. It may also be superior to wire rope in low-temperature applications and in areas where there are specific chemicals that are corrosive to steel products. Another significant advantage is that synthetic rope is roughly 80 percent lighter than wire rope. It also has no tendency to twist, since no significant torque, or twisting action, is applied to the material during the manufacturing process. Synthetic rope does not require periodic lubrication. The material is also nonconductive, making it safer to use around power distribution and transmission systems as long as it is dry. Utility companies sometimes use it on digger derricks for this reason.

On the other hand, synthetic rope may be damaged by exposure to the sun's ultraviolet (UV) rays. UV rays degrade many synthetic materials, including many plastics, causing them to harden and/or become brittle. Dirt and grit that works its way into the rope creates additional friction and subsequent damage during use. For longevity, it must be kept very clean. Synthetic wire rope is also not recommended for use above the 140°F (66°C) range. It is unknown if ASME will broadly accept the material in the future, and the cost is well above that of comparable wire rope. It is important to note that every mobile crane application must be approved by the manufacturer. Still, it is clear the technology is sound and future developments may eventually lead to synthetic materials replacing wire rope in a variety of applications.

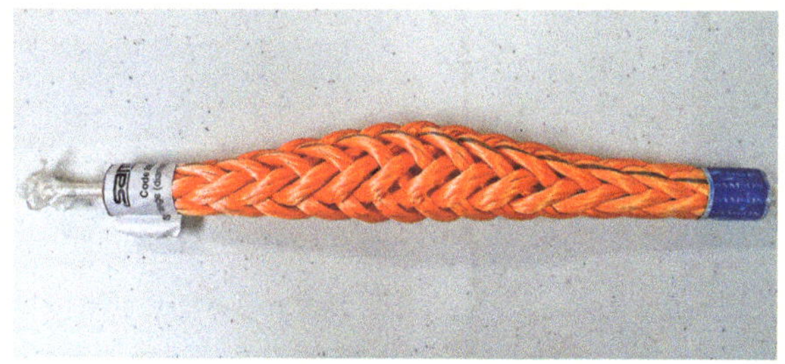

1.1.1 Wire Rope Component Arrangement

Wire rope component arrangement refers to the pattern of the components of wire rope. This is important in order to identify the characteristics of a given wire rope and its possible uses. Users must first understand the directional relationship between the strands and the core of the rope. This relationship or pattern is known as the lay of the rope. The word *lay* is used in several ways relative to the construction of wire rope, explained as follows:

- The direction of the strands is referred to as either a *right lay* or *left lay*. Right lay indicates that the wrap of the strand is in a clockwise direction around the core. Left lay indicates that the direction of wrap is in the counterclockwise direction. In other words, when holding the rope pointing straight ahead and away from you, when the strands turn to the right, it has a right lay. When they turn to the left, it has a left lay.
- The terms *regular lay*, *lang lay*, and *alternate lay* refer to the strand-to-strand relationship. Regular lay indicates that the individual wires of each strand run parallel to the direction of the run of the rope. *Figure 2* shows examples of both right regular lay (RRL) and left regular lay (LRL) wire

ropes. When the rope has a lang lay, the individual wires in each strand are oriented in the same direction as the strand is oriented in the rope. *Figure 3* shows examples of right lang lay (RLL) and left lang lay (LLL) wire ropes. Comparing them to the ropes shown in *Figure 2* helps to distinguish one style from the other. Wire rope with an alternate lay combines the features of both regular lay and lang lay wire ropes by alternating strand arrangements. A right alternate lay (RAL) wire rope is shown in *Figure 4*. Alternate lay rope is available in both right and left configurations.

- The length of each lay is also a factor in wire rope construction. The lay length is the distance traveled as each strand makes a complete revolution around the core, as shown in *Figure 5*. The length of the lay is often an important factor during wire rope inspections, where some maximum number of broken wires may be allowed per lay length.

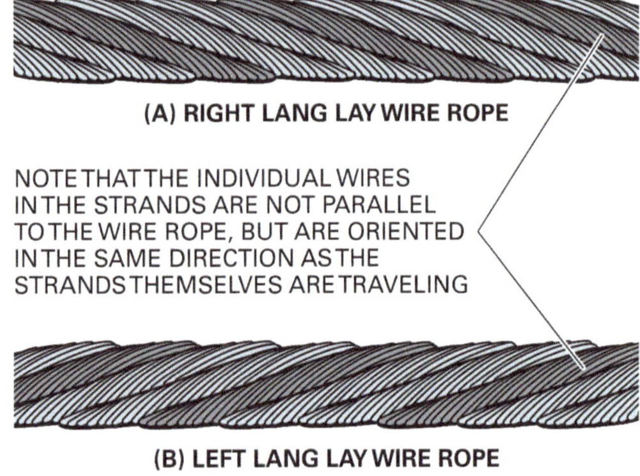

Figure 3 Right lang lay and left lang lay wire ropes.

Figure 4 Right alternate lay wire rope.

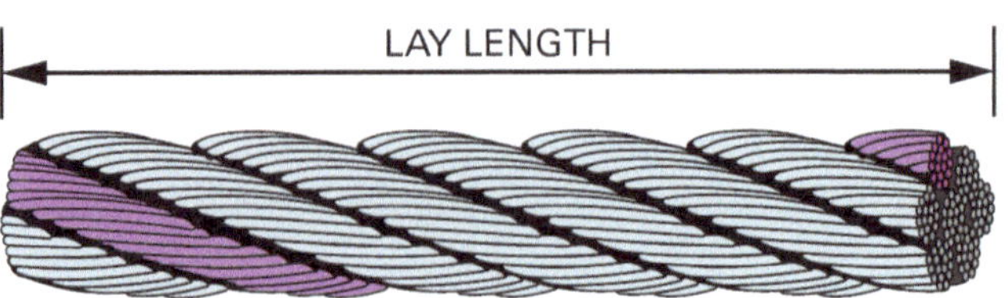

Figure 5 Determining lay length.

Figure 2 Right regular lay and left regular lay wire ropes.

1.1.2 Comparison of Characteristics

Regular lay ropes are easier to handle, resist kinking and twisting, and are relatively stable. For this reason, right regular lay rope is the most commonly used wire rope on mobile cranes for hoisting, and for a number of other applications as well. Lang lay ropes also have some advantages. For example, they resist fatigue better when bent around a sheave. They are also said to be more resistant to abrasion.

However, lang lay rope has significant disadvantages when it comes to crane applications. Lang lay rope tends to rotate, or unwind, when either end of the cable is not rigidly fixed and a load is imposed. As it unwinds, the load is progressively transferred to the core of the rope, creating the potential for failure. Another concern is that it does not resist crushing on the hoist drum as well as regular lay rope.

Alternate lay rope combines the features of regular and lang lay ropes. It offers the advantages of both constructions while minimizing the disadvantages. Although right regular lay rope is the most common type found on cranes, there are also applications served best by alternate lay ropes.

1.1.3 Strand Patterns

Wire rope is comprised of a number of strands wrapped around a core, and each strand is comprised of wires. The wires in the strands can be arranged in a variety of ways in order to achieve specific characteristics.

There are four basic patterns used to create strands for a wire rope. *Figure 6* shows the four basic strand patterns: the wire, Warrington, Seale, and filler wire configurations. Note that a number is shown beside each of the different patterns. The number indicates how many wires are used to create the strand. Once the strands are combined around a core to create a wire rope, another number is added to indicate the number of strands used in the rope. For example, consider a rope that is constructed using (6) 19 Warrington strands. This wire rope is designated as 6 × 19 rope.

These four basic patterns may be combined to create a different strand. *Figure 7* shows a 49 Warrington Seale pattern, which combines the Warrington and the Seale patterns. Starting from the inside, the first two layers conform to the Seale pattern. The third layer conforms to the Warrington pattern, while the outer layer returns to the Seale pattern.

Another characteristic of wire rope is related to the forming of the wire and/or strands. The wires forming a strand can be straight, as manufactured, and then twisted as the strand is formed. This type of construction is considered non-preformed. However, the individual wires that will be used to construct the strand can also be twisted, before the strand is formed, into the exact helical shape they will maintain as part of the strand. This process uses preformed wires. Note that these terms can also be applied to the final wire rope assembly. For example, a strand constructed using preformed wire is referred to as a preformed strand. Preformed wire rope is constructed of strands that were preformed before the rope was constructed, and those strands were also preformed.

One difference between the two types can be seen when a wire breaks in a strand. A broken wire in a non-preformed strand will likely protrude from the bundle significantly. If the strand is preformed, the broken wire will not protrude as significantly from the bundle. Preformed wire rope is the type most often used on cranes.

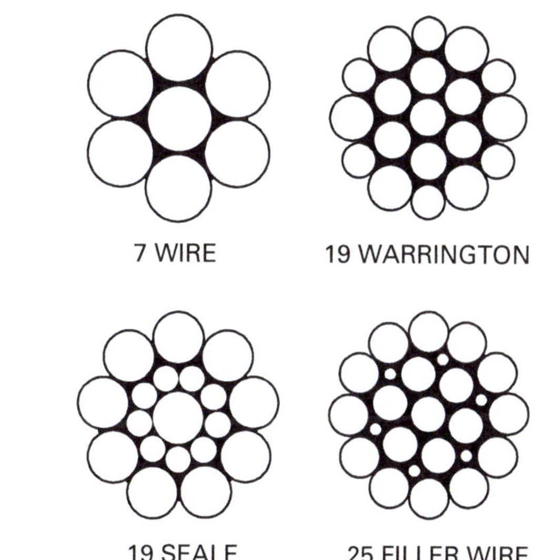

Figure 6 The four basic strand patterns.

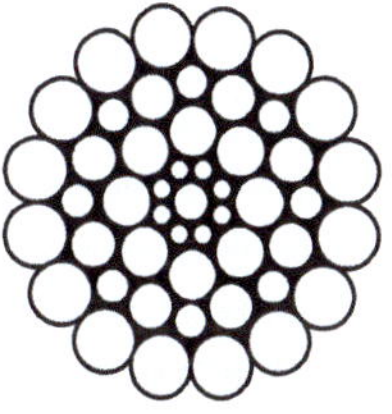

Figure 7 A strand constructed by combining the Warrington and Seale patterns.

1.1.4 Rotation-Resistant Wire Rope

Due to the twist imposed on wire rope when it is manufactured, there is a tendency to twist in the opposite direction of the lay when it is under load. This can create significant problems. Ideally, a wire rope would have no tendency to twist at all.

Rotation-resistant wire rope is designed to serve that purpose. Note, however, that rotation-resistant rope is not rotation-proof. The difference in construction is found in the lay of the inner and outer strands (*Figure 8*). In non-rotation-resistant ropes, the lay of the inner strands follows the lay of the outer strands, so that the torque of the strand layers is working in the same direction. In rotation-resistant rope, the inner and outer strands have an opposing lay, so that the torque of the strand layers oppose each other. The result is a significantly reduced tendency to twist under load. This type of rope is especially preferred for single-reeved load lines such as whip lines.

The advantage of rotation-resistance comes at a cost, however. These ropes are somewhat more fragile and must be handled with greater care. There are also different OSHA standards applied to them during an inspection. With the lay of the strands in opposing directions, greater wear should be expected between the core and the adjacent strand layer. They are generally more susceptible to kinking, crushing, and birdcaging. When compared to equal sizes of other types of wire rope, rotation-resistant wire rope has a lower

capacity. It is important to remember that they are not rotation-proof—some rotation will likely be experienced, especially with new rope.

There are three types of rotation-resistant wire rope defined in 29 *CFR* 1926.1414(e):

- *Type 1* – Ropes that have at least 15 outer strands, and three layers of strands over a core. They have the greatest resistance to rotation.
- *Type 2* – Rope with 10 or more outer strands and two or more layers of strands over a core. They are considered to have significant resistance to rotation.
- *Type 3* – Ropes that have no more than 9 outer strands with two layers of strands over a core.

There are a number of regulations contained in 29 *CFR* 1926.1414(e) related to rotation-resistant rope. Some are presented in other sections of this module. Note that certain portions of *ASME Standard B30.5, Section 5-1.3.2* have been incorporated by reference into the OSHA standards. Refer to the OSHA standard for details of the incorporation.

1.1.5 Measuring Wire Rope

All ropes, whether wire or fiber, are measured by the diameter of the circle that fully encloses the rope cross section. Refer to *Figure 9*. Another way to explain the measurement is to say that the largest measurement that can be made on a section of wire rope represents the correct measurement.

Note that all new wire rope is made slightly oversized, as it will attain its design size during the break-in period. This can lead to a certain amount of confusion when measuring a rope. The rope may also measure somewhat undersized as it approaches the end of its life, due to stretching and tightening of the wires and strands. Although replacement criteria are discussed later in this module, a reduction of 5 percent in the diameter of a rope is one cause for removal and replacement.

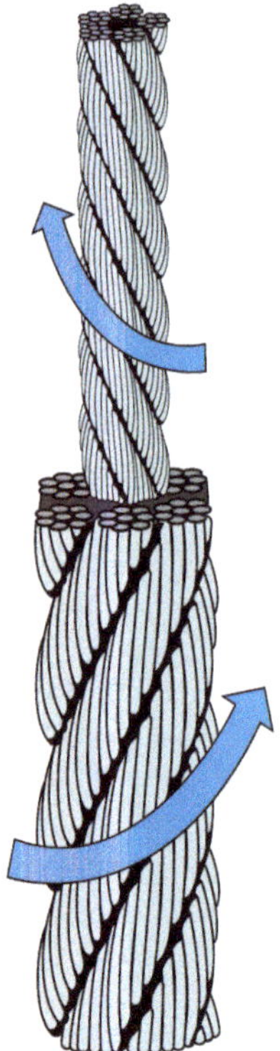

Figure 8 Rotation-resistant wire rope construction.

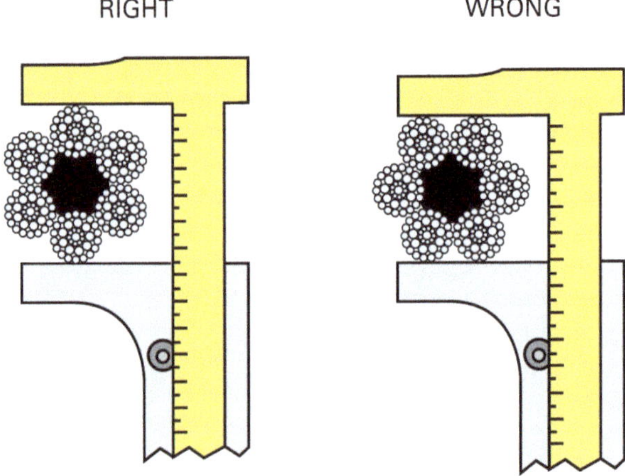

Figure 9 Measuring the diameter of wire rope with a caliper.

Wire Rope in a Metric World

Most of the wire rope industry is making a slow transition to the metric system. There are several characteristics of wire rope that must be expressed in different units. The three primary characteristics requiring conversion are diameter, weight, and breaking force. The industry has agreed on what is referred to as a "soft conversion" to the metric system of measurement in these areas for the foreseeable future.

For diameter, the English units of fractional and whole number diameters are converted to a metric size by multiplying inches by 25.4, which is the conversion factor to millimeters (mm). Then the metric size closest to the result is identified as the correct equal in size. For example, 1" rope converts to 25.4 mm in size. In this range, metric wire rope offers diameters of 22 mm, 26 mm, and 29 mm. So the size considered equivalent to 1" is the 26 mm size.

The weight of wire rope is documented in pounds per foot, or lbs/ft. This is converted to the metric unit of kilograms per meter (kg/m) by multiplying by 1.488.

The breaking force is typically documented in tons (T). Metric wire rope is rated in kilonewtons (kN) or kilograms. To convert tons to kilonewtons, multiply tons by 8.897. To convert tons to kilograms, multiply tons by 907.2.

1.1.6 Wire Rope Product Designations

This section explains how to identify products based on the various characteristics of wire rope construction. The identifying information for wire rope addresses the following characteristics:

- Length
- Diameter
- Finish
- Lay
- Grade
- Core type
- Preformed (Pref) or non-preformed (Non-Pref).

For example, the description of a wire rope product may read as follows:

1,200 feet $\frac{3}{4}$" 6 × 19 WS Pref RRL EIPS IWRC

This description indicates that the product is 1,200 feet (365 meters) of $\frac{3}{4}$" (19.05 mm) diameter rope. The rope is constructed using six strands around a core, with each strand constructed in a 19 Warrington Seale pattern. Therefore, each strand has 19 wires. The rope is preformed; it has a right regular lay (RRL); is made using extra improved plow steel (EIPS); and is built around an independent wire rope core (IWRC).

Although users such as crane operators and riggers are not usually responsible for ordering wire rope, it is still helpful to understand that not all wire rope created under a particular classification—6 × 19 wire rope, for example—does not necessarily consist of strands with 19 wires each. Wire rope classified as 6 × 19 may actually have as many as 26 wires in the strand. *Figure 10* shows three different wire ropes that all belong to the 6 × 19 class of wire ropes.

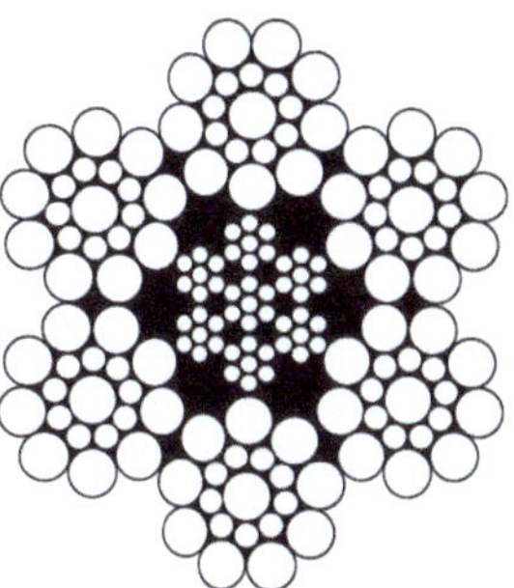

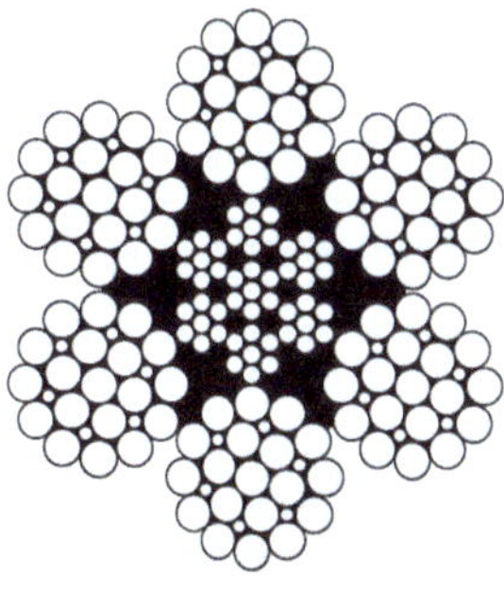

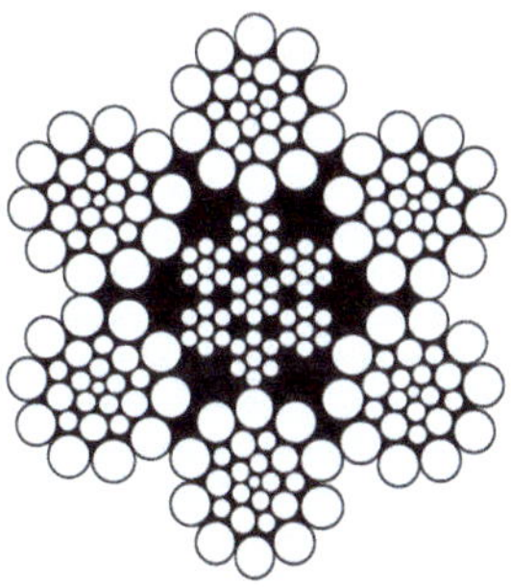

Figure 10 Examples of wire ropes in the 6 × 19 class.

For this reason, the number of strands shown in a wire rope description may vary unless an order is placed with a specific request for strand count. The reason this is allowed in the industry is because all of the wire ropes placed in the 6 × 19 class, regardless of the actual number of wires per strand, share the same weight per foot and minimum breaking strength. However, the differences in wire count per strand has an effect on the functional properties of the rope. Other wire rope classes, such as the 6 × 36 class, also include products with varying wire counts per strand.

1.2.0 Terminations

Both ends of a wire rope must be secured in some manner. On a mobile crane's hoist rope, one end is normally secured to the drum, while the other is affixed to a point on the load block. This enables the power of the machine to be transmitted to the load for lifting purposes. It is important to understand that terminations generally represent the weakest link, and a poorly made termination is a definite liability.

The wire rope ends require preparation to enable it to be reeved through the sheaves and then connected at the drum. Terminations also often require the same preparation. The end may be welded and tapered, or seizing may be used to capture the wires and strands. Both of these preparations are shown in *Figure 11*. Manufacturers may also fuse the end of a wire rope, using current passed through the end to fuse the individual wires together, rather than using a traditional welding method. The preferred method will generally be specified by the wire rope manufacturer. After the end preparation is completed, the rope can then be threaded through the narrow

Figure 11 End preparations.

opening between the sheaves and guards and the drum connection point without the wires or strands unravelling.

The other end of the wire rope must also be secured at the load block or ball. As shown in *Figure 12*, there are a number of different methods used to accomplish this, including the following:

- Swaged socket
- Poured socket
- Thimble with a mechanical splice
- Thimble with wire rope clips
- Thimble with a hand-tucked splice
- Wedge socket

As shown in *Figure 12*, each of these terminations has an efficiency value, expressed as a percentage. The value relates to the minimum breaking strength of the wire rope. For example, if the termination is a poured socket, which has an efficiency rating of 100 percent, this means the termination is as strong as the connected wire rope. Terminations that have percentages below 100 percent are not as strong as the wire rope itself, and therefore represent a weaker link in the assembly.

A very popular termination, the wedge socket, has an efficiency of 75 to 80 percent, depending on the product. This means that the wedge socket has 75 to 80 percent of the wire rope's minimum breaking strength. This termination is therefore the likely point of failure if all the components of a wire rope assembly are stressed to capacity. However, it is not a given that it will fail first.

1.2.1 Fabricated Sockets

There are two types of fabricated sockets: poured and swaged. Poured sockets consist of a metal cup or cone with a termination such as a thimble attached. The wire rope is attached to the socket with liquefied zinc or a zinc alloy that is poured into the cup through which the wire rope is inserted. When the metal cools and solidifies, it holds the rope firmly in place. The molten metal connection is commonly referred to as a speltered socket. A variation on this type of termination replaces the metal with an epoxy resin that hardens through a chemical reaction, holding the wire rope in place. The efficiency of both types is 100 percent of the rope's minimum breaking strength. The selection of which material to use depends on the application. Note that the epoxy resin is vulnerable to excessive heat or chemically caustic atmospheres.

Swaged sockets are visually similar to poured sockets. The difference is in how the wire rope is secured. A swaged socket secures the wire rope mechanically. Swaging is a process in which the shank of the fitting is pressed or crimped tightly to the wire rope, forming a mechanical bond. If the fitting is properly pressed and the wire rope properly prepared, this type of termination has a 100-percent efficiency rating. However, this is based on using IWRC wire rope. This type of fitting is not recommended for use with fiber core wire rope.

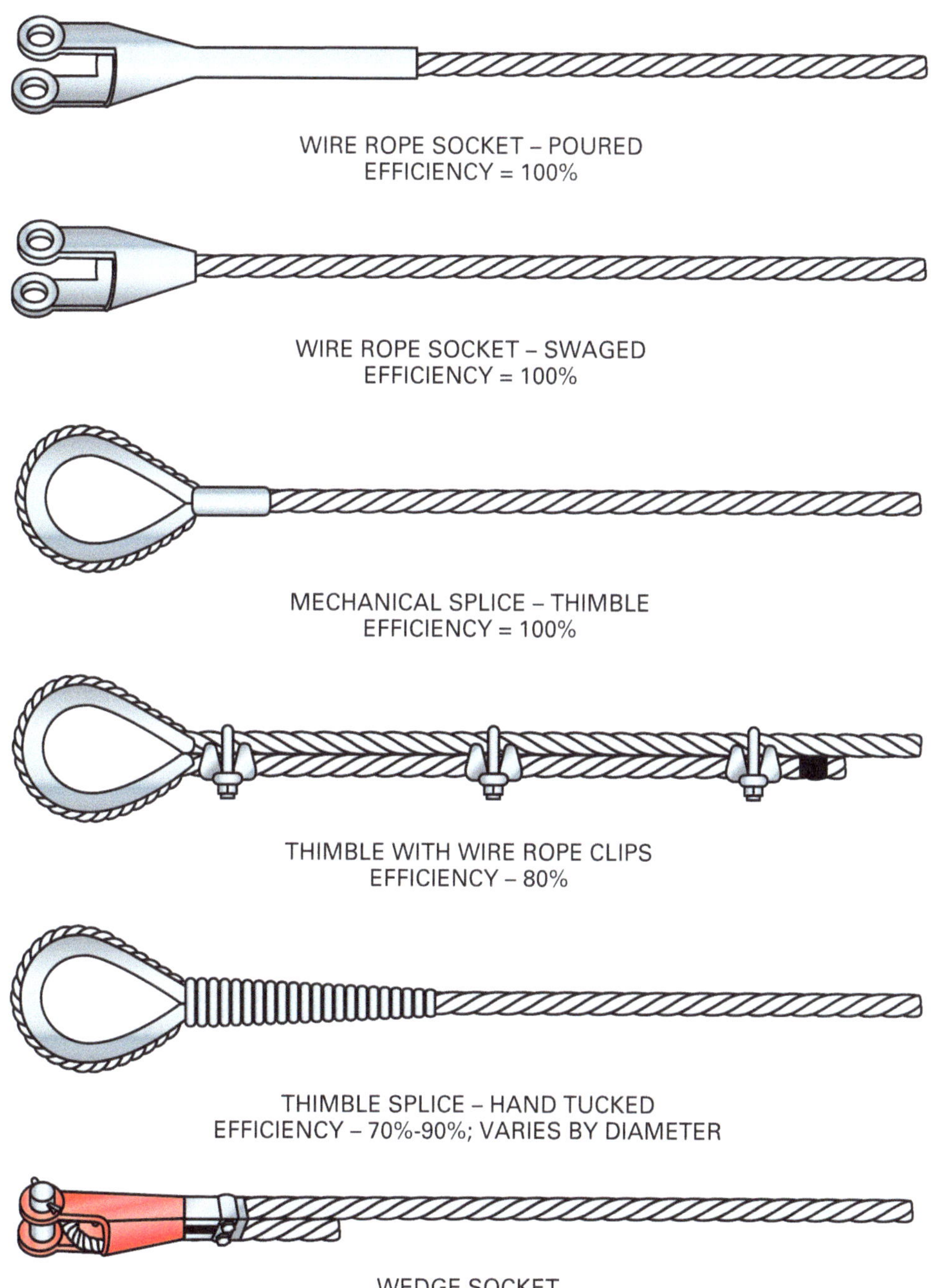

Figure 12 Examples of end fittings.

1.2.2 Wedge Sockets

The wedge socket (*Figure 13*) is a very popular termination. There are several different styles, but all are similar in concept. Wire rope is first threaded through the socket. A wedge is then placed into the loop formed and pushed back into the socket using a jackbolt. As a load is applied, the wedge seats tighter into the socket, holding the wire rope in place and becoming tighter as the load increases. The shank of the wedge socket has a removable pin affixed. It is used for mounting a load block or other component.

The socket, wedge, and pin should be inspected for defects and damage before use. Do not use parts that are corroded, cracked, or worn. The assembly must also be of the correct size for the wire rope. Consult the socket manufacturer for recommendations regarding the specific use and reapplication of wedge sockets as necessary.

> **CAUTION**
>
> Careful attention should be given to wedge socket components. The components are a matched set and should be kept together as an assembly. Never randomly assemble them from spare parts.

To understand termination assembly, you must know how to determine the live section of a wire rope from the dead end. This concept is easily explained by the following: the live section of a wire rope is the section that will endure the load. The weight of the load is not applied to the dead end.

When assembling a wedge socket, there are important checkpoints to keep in mind to prevent incorrect assembly:

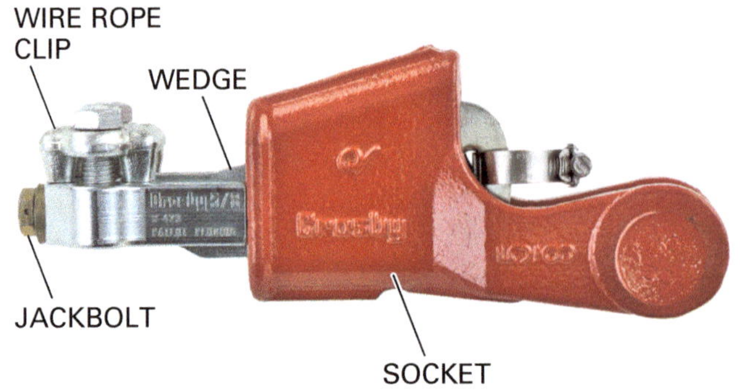

Figure 13 Example of a wedge socket assembly.

- Do not mix and match wedge socket components of different sizes or model numbers.
- Do not attach the dead end to the live section of the wire rope.
- Secure the dead end of the wire rope.
- Align the live section of the wire rope with the center line of the pin.
- Exercise care when seating the wedge into the socket. Follow the manufacturer's recommendations for seating the wedge before use.
- The tail length of the dead end of wire rope should be a minimum of six rope diameters but not less than 6".

Figure 14 shows that there are correct and incorrect methods of positioning the wire rope, as well as completing the termination of the dead end. Note that this figure demonstrates the correct and incorrect assembly of just one specific type of wedge socket. It is essential that the manufacturer's instructions for each specific type be consulted. The section of wire rope in line with the pin will experience the load weight; therefore,

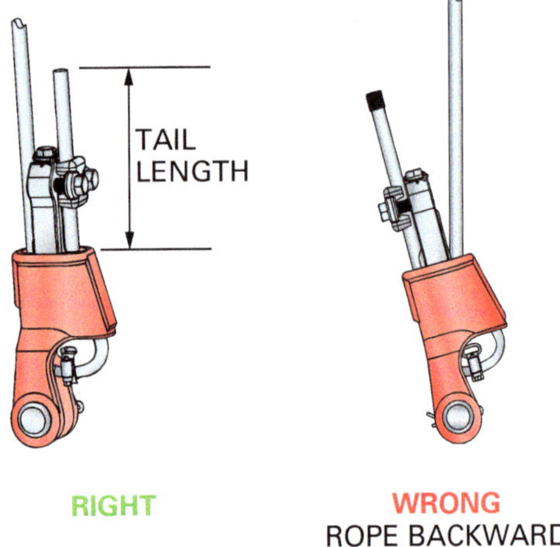

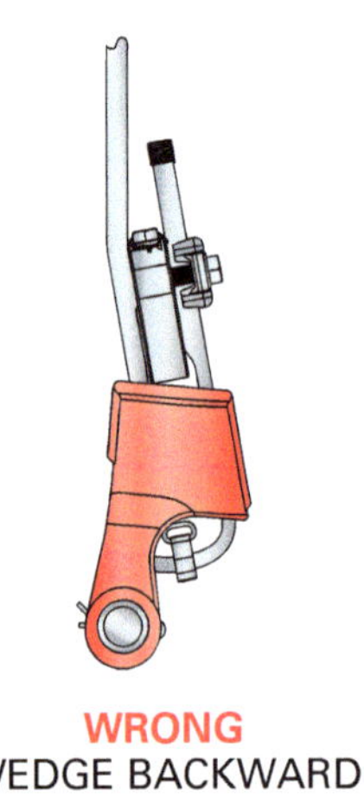

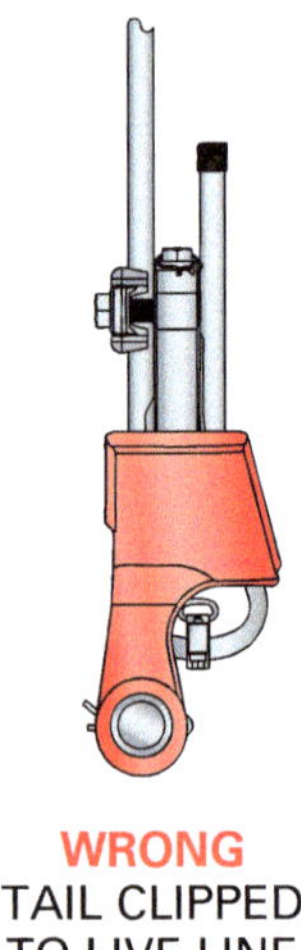

Figure 14 Wedge socket assembly.

NCCER – *Intermediate Rigger*

it is considered the live section of the rope. In contrast, the end that wraps around the wedge and exits the socket is considered the dead end, which does not experience the weight of the load. Securing the dead end of the wire rope to the wedge with the built-in wire clip prevents it from slipping out of the socket in an unloaded condition. The threaded parts of the wedge socket have torque values specified by the manufacturer that must be applied

1.2.3 Wire Rope Clips

Wire rope clips are used to create terminations with thimbles and to terminate wedge socket assemblies. There are two common designs for wire rope clips: the forged U-bolt and the fist grip (*Figure 15*). The most obvious difference between the two is that the fist grip does not have a U-bolt. Instead, it has two saddles that bolt together. The U-bolt clip has two U-bolt nuts on the same side of the clip, while the fist grip has mounting nuts on opposing sides.

The U-bolt style of clip consists of the U-bolt, the nuts, and the saddle. To properly mount U-bolt wire-rope clips, you must first identify the live section and dead end of the rope. The direction that the U-bolt and the saddle face is critical to achieving full strength for the application.

In all applications, the saddle of a U-bolt clip is positioned on the live section of wire rope, and the U-bolt is against the dead end (*Figure 16*). The design of the saddle provides protection for the live section of rope, preventing crushing and

deformation when the U-bolt nuts are tightened during the installation.

The fist grip clip has the unique advantage of having no right or wrong direction for mounting. The two sides are mirror images of each other, having a saddle on each, which protects the wire rope when the mounting bolts are tightened.

The number of clips needed, the torque applied to the bolts, and the length allowed for the dead end are all specified by the clip manufacturer based on rope size. A common practice when using more than one wire rope clip is to space them evenly. The nuts should also be tightened evenly and alternately with a torque wrench until they reach the required torque. It is mandatory to test-load the assembly. The test load weight should be equal to or greater than the weight of the expected load. After testing, the nuts should again be tightened to ensure the specified torque remains. Like wedge sockets, the manufacturer's specifications should always be followed for installing and testing wire rope clips.

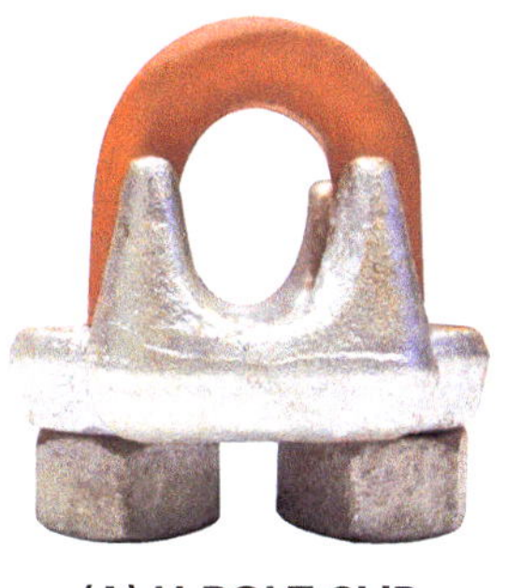

(A) U-BOLT CLIP   **(B) FIST-GRIP CLIP**

Figure 15 Wire rope clips.

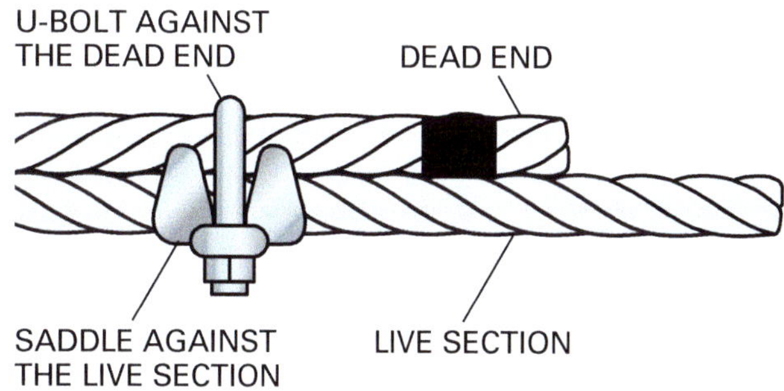

Figure 16 Assembling a U-bolt wire rope clip.

Additional Resources

ASME Standard B30.5, Mobile and Locomotive Cranes. Current edition. New York, NY: American Society of Mechanical Engineers.

ASME Standard B30.9, Slings. Current edition. New York, NY: American Society of Mechanical Engineers.

29 *CFR* 1926, Subpart CC, *Cranes and Derricks in Construction.* Current edition. Washington, DC: US Department of Labor, Occupational Safety and Health Administration.

29 *CFR* 1926.251, *Rigging Equipment for Material Handling.* Current edition. Washington, DC: US Department of Labor, Occupational Safety and Health Administration.

Mobile Crane Safety Manual. 2014. Milwaukee, WI: Association of Equipment Manufacturers.

Willy's Signal Person and Master Rigger Handbook. Current edition. Ted L. Blanton, Sr. Altamonte Springs, FL: NorAm Productions, Inc.

North American Crane Bureau, Inc. website offers resources for products and training, **www.cranesafe.com**.

Union®, a WireCo® WorldGroup brand, **www.unionrope.com**.

1.0.0 Section Review

1. What type of lay is indicated when the strands are wrapped around the core to the right but the outer wires in each strand are running parallel to the wire rope itself?

   a. Right regular lay
   b. Left regular lay
   c. Right lang lay
   d. Alternate lang lay

2. When a wedge socket or similar termination is said to have an efficiency of 80 percent, it means that _____.

   a. it will hold its rated load roughly 80 percent of the time
   b. the socket is an appropriate choice about 80 percent of the time
   c. the completed assembly has 80 percent of the rope's minimum breaking strength
   d. the termination can be used on roughly 80 percent of the available wire rope types

2.0.0 WIRE ROPE INSPECTION, HANDLING, AND MAINTENANCE

Objective

Identify and describe wire rope inspection, handling, and maintenance requirements.

a. Identify standards-based wire rope inspection requirements and the actions to be taken for various deficiencies.
b. Describe the various types of wire rope deficiencies that may be encountered.
c. Describe how to inspect sheaves and load blocks.
d. Describe common wire rope handling and maintenance procedures.

Performance Tasks

2. Inspect wire rope using the appropriate inspection criteria.
3. Inspect a sheave and load block using the appropriate inspection criteria.

Trade Terms

Constructional stretch: The stretching that occurs in wire rope when it is initially placed in service, during its break-in period. Constructional stretch is expected in all wire ropes. Constructional stretch in standard ropes is generally 0.25 to 1 percent of the rope's length.

Crossover points: Per 29 *CFR* 1926.1401, specific locations where a wire rope layer spooled on a drum must climb up and cross over the previous layer. Crossover points are located at each end of the drum at its flanges, where the rope must change direction to continue spooling onto the drum.

Fatigue fractures: Progressive fractures resulting from the repeated bending of individual wires. These fractures may occur at bending-stress levels well below the ultimate strength of the material. Lack of lubrication and rust on a wire rope can accelerate fatigue fractures.

Flange points: Per 29 *CFR* 1926.1401, points where the wire rope contacts the drum flanges at each end.

Pendants: Fixed lengths of rope with mechanical fittings at each end.

Repetitive pickup points: Per 29 *CFR* 1926.1401, refers to the sections of a rope where short-cycle operations cause it to be repeatedly spooled on and off a small portion of the drum.

Running ropes: Wire ropes that move over sheaves and/or drums.

A well-planned and thoroughly documented program of periodic wire rope inspection and maintenance is required to meet regulatory requirements and to ensure that the rope will be safe for use and provide maximum service life. An experienced inspector must be able to determine the cause of various rope deficiencies and make judgments on the continued service life of the wire rope based on observations.

2.1.0 Standards-Based Inspection Requirements

A good inspection program will provide guidelines to inspectors that aid in determining the condition of a wire rope and its suitability for continued service. The following information and guidance is based on the standards of both the Occupational Safety and Health Administration (OSHA) and the American Society of Mechanical Engineers (ASME).

2.1.1 OSHA Wire Rope Inspections

29 *CFR* 1926.1413 covers the requirements for wire rope inspection on mobile cranes in the construction environment. OSHA uses the following terminology to identify wire rope inspections, which is the same terminology used to describe crane inspections:

- *Shift inspection* – An inspection required prior to each shift of crane operation. Note that this is not necessarily done once a day. If the crane operates three shifts during the course of a day, then three shift inspections are required.
- *Monthly inspection* – As stated, an inspection that must be accomplished at least once each month.
- *Annual/comprehensive inspection* – An inspection that must be accomplished at least once every 12 months.

During all three inspections, the inspector is looking for the same deficiencies in the wire rope. Deficiencies are placed into one of three categories. The category of the deficiency determines what action must be taken next. The primary difference between the three periodic inspections is the effort made to access all sections of the wire rope.

During shift inspections, the portion of the wire rope that is likely to be in active use during the shift must be visually inspected. However, OSHA does not require that the boom be lowered or that a great deal of wire rope be unwound from the drum. OSHA does not require specific documentation of the wire rope shift inspection, but it may be required by the employer or crane owner.

The monthly inspection required by OSHA is virtually the same as the shift inspection. There are two primary differences. The monthly inspection must be clearly documented, and specific attention must be given to any issues that were identified for monitoring on a previous annual inspection. The qualified person that accomplishes the annual inspection may identify one or more issues that are not immediate safety hazards, and document those issues for monitoring over the course of the next year. These issues must then be rechecked during each monthly inspection.

Annual inspections also include all the inspection points of the shift inspection. However, there are some differences in the effort required. During the annual inspection, the entire length of the wire rope must be inspected. Extra attention is given to the portion of the rope that is normally hidden from view during shift and monthly inspections, and to those areas of the rope that commonly pass over sheaves.

One additional requirement found in the OSHA standards is that wire rope lubricants that interfere with visual inspection of the rope must not be used.

2.1.2 ASME Wire Rope Inspections

ASME also uses the same inspection terminology applied to crane inspections to identify wire rope inspections. Per *ASME Standard B30.5, Section 5-2.4.2*, frequent inspections are those conducted each working day. The frequency of wire rope periodic inspections are left to be determined by a qualified person. The ASME standard indicates that the individual should determine the frequency based on factors such as expected rope life, the operating environment, frequency rates of lifting operations, and the percentage of lifts made at or near crane capacity. Inspections do not need to be made in equal increments of time, but the frequency of inspection should increase as the rope nears the end of its working life. The only stipulation is that the periodic inspection of wire rope must occur at least annually.

2.1.3 OSHA Wire Rope Deficiency Categories

Since the OSHA standard carries the weight of law, the inspection criteria outlined in 29 *CFR* 1926.1413 will be the primary focus of this section. *ASME Standard B30.5* does not deviate significantly in the criteria. The wording is virtually the same in some areas.

As noted earlier, OSHA places wire rope deficiencies into specific categories. The category of a deficiency determines the action to be taken. The three categories and the related deficiencies are identified here. Descriptions and examples of various wire rope deficiencies are presented in the next section.

When a Category I deficiency is identified, a competent person must determine if it is a safety hazard. If it is a safety hazard, then the wire rope cannot be used until it is either replaced completely, or the rope is cut and the deficient portion removed. Cutting the rope is acceptable as long as enough rope remains. The drum must have at least two wraps of rope remaining when the boom and/or the load is in its lowest position.

Damaged portions of wire rope can be cut off and removed. However, the rope cannot be spliced in any way to increase its length. Splicing the rope at any point along its length is strictly prohibited.

Category I deficiencies include the following:

- Significant distortion of the wire rope, including kinks, crushed sections, birdcaging, signs of core failure, and separation of strands
- Significant corrosion
- Damage from an electric arc, such as from nearby welding (note that this does not include arc damage resulting from power line contact)
- Improperly assembled or misapplied terminations
- Corroded, cracked, or bent terminations, or terminations suffering from excessive wear

Category II deficiencies are more significant than those in Category I. When a Category II deficiency is identified, the owner or employer still has the option of replacing the wire rope completely or cutting off the defective portion. However, the OSHA standard also allows the owner to consult with the wire rope manufacturer and comply with its guidelines for rope replacement. In other words, the rope can remain in service as long as the damage does not exceed the manufacturer's guidelines for removal. Category II deficiencies include the following:

- Broken wires; there are specific allowances for broken wires that are dependent on the type of rope in use and/or the application.
  - *For running ropes*: The rope is deficient if it has six broken wires randomly distributed in one lay length, or three broken wires in the same strand of one lay length. Remember that a lay length is the distance required for a single strand to make a complete rotation around the rope assembly.
  - *For rotation-resistant ropes*: The rope is deficient if there are two broken wires randomly distributed along a length of six rope diameters. For example, on a 1" (2.5 cm) rope, two broken wires found within 6" (15.2 cm) of each other would be cause for rejection. It is also deficient if there are four broken wires randomly distributed along a length of 30 rope diameters.
  - *For standing wire ropes*/pendants: The rope is deficient if there are more than two broken wires in one lay length in portions of the rope away from any terminations. If it is at a termination, more than one broken wire in a lay length is not allowed.
- A reduction of the rope diameter that exceeds 5 percent of the rope diameter. This can occur as the result of abrasion, stretching, or crushing.

Category III deficiencies are considered the most significant. There is no provision for the manufacturer to allow the rope's continued use. The rope must either be replaced or the defective portion cut off and removed. Category III deficiencies include the following:

- When rotation-resistant rope is in use, a protrusion of the core through the side of the rope, or any other evidence of core failure that may be encountered
- Evidence of contact with an energized power line
- A broken strand

The OSHA standard also identifies certain items that should be given special attention during an inspection of the wire rope. These are referred to as critical review items in the standard. They include:

- Rotation-resistant wire rope
- Rope used on boom and luffing hoists, especially where reverse bends occur
- Rope at flange points, crossover points, and repetitive pickup points on the drum
- Rope at or in the vicinity of terminations
- Rope that contacts saddles or sheaves where the rope may travel back and forth repeatedly over a limited length of rope

2.2.0 Deficiencies in Wire Rope

To effectively inspect wire rope, users must understand what constitutes damage and/or a defect. There are many different types of damage that can occur. A great deal of research has enabled rope manufacturers to not only identify different types of damage, but also to identify how and why the damage occurred with surprising precision. Although the scope of this module does not allow for a complete study of the topic, a number of common wire rope deficiencies are described as follows (refer to *Figure 17*)

- *Broken wires* – The number of broken wires in a wire rope is an indication of its general condition. In many cases, these failures are related to bending or vibration in sheaved or reeved areas of the wire rope. Criteria have been established to mandate wire rope replacement based on the number of wires in the rope that are broken and where those breaks occur. (The OSHA criteria for this Category II defect were presented in the previous section.) Once broken wires start to appear in a wire rope, it is prudent to assume that more will appear in a short time under normal operating conditions. Note that a close examination of the two ends of a broken wire will reveal the likely source of the breakage. A wire break that results in both ends having a slight tapered appearance, and also results in one tip being cup-shaped (concave) while the other is conical (convex) is likely the result of loads that exceeded its tensile strength. In other words, the wire was stretched to the point of failure. If the wire ends are consistent in diameter and the tips are flat and even, the break likely happened due to simple fatigue. When broken wires with these characteristics are found in groups at the crown (top) of the strands or in the valleys between strands, they are referred to as fatigue fractures or fatigue breaks. Refer to *Figure 18*.
- *Kinks* – Kinks are permanent distortions caused by loops in the rope having been drawn too tightly. Ropes with kinks must be removed from service. To avoid kinks, never allow them to be twisted significantly. When twisted rope goes slack, a loop can form easily.
- *Birdcaging* – Birdcaging is caused by mistreatment of the wire rope, such as sudden stops of the load that allow it to bounce up and down, the rope being pulled through tight sheaves, or the rope wound on too small a drum. This is cause for rope replacement or removal of the damaged section.

- *Crushing* – The first layer of rope on a hoist drum is most likely to experience crushing, as the stress is applied to all the layers above. Crushing at points where the rope crosses over itself often leads to broken wires.
- *Corrosion* – Corrosion can cause serious degradation and weakening of a wire rope. Wire ropes require the application of lubricants that help prevent corrosion. Wire rope corrosion can occur internally before there is any external indication, since water can be trapped there for longer periods of time, and it is more difficult to force lubricants into the core. The presence of corrosive pitting is often grounds for immediate removal of the rope from service.

There are other forms of damage that wire rope may suffer. It may be difficult to determine how significant a defect is from a safety point of view. Whenever there is doubt about the integrity of a wire rope, consult the rope manufacturer directly and provide pictures to help them assess the problem. The following are other types of damage to look for during an inspection of the wire rope:

- *Abrasion* – Abrasion occurs when wire rope moves across or through any abrasive medium. Abrasion can occur, for example, as the rope runs over sheaves and drums. Abrasion also occurs internally as the different strand layers move in opposing directions. During normal operation, this wear is not visible. Excessive vibration or whip can cause both abrasion and fatigue. Drum crossover and flange point areas must be examined closely. If the effects of abrasion result in the diameter being reduced more than 5 percent, it becomes a Category II defect.
- *Electric arc* – An electric arc occurs when a wire rope contacts an energized power source. There have been cases of welders using the rope as a ground for an arc-welding task. This can lead to arcs occurring at various places along the length of the rope. As electricity passes through the wire rope, some of the wires might be fused, annealed, and/or discolored. A significant arc will appear as a melted spot in the rope, often accompanied by metal splatter around the area. 29 *CFR* 1926.1413 requires the removal of wire rope that has been damaged from contact with an energized power line.
- *Heat damage* – High temperatures may cause metal discoloration and drive lubricants out of the rope. Fiber core ropes are particularly vulnerable to excessive heat.

Figure 17 Examples of wire rope failure.

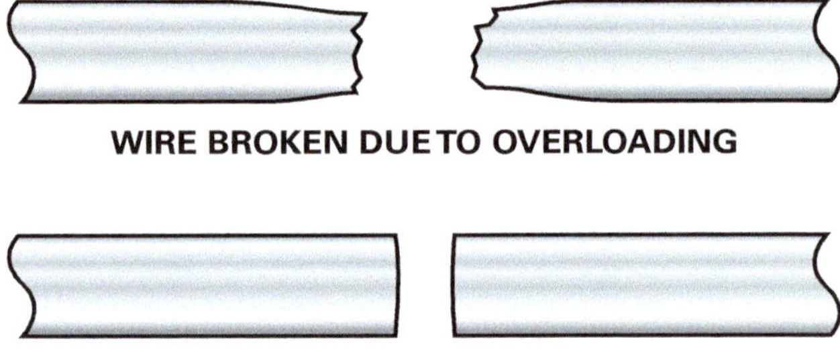

Figure 18 Examples of wire breaks.

- *Protruding core* – Core protrusion, as shown in *Figure 19*, occurs when the core visibly protrudes through one or more strand layers. Core protrusion, as well as broken cores, is often caused by repetitive shock loading of the rope.
- *Rope stretch* – All rope will initially stretch as load is applied. Wire rope stretches in three phases. During the first phase, when the rope is new, it will stretch as loads are applied and the wires and strands are all pulled tighter against each other. This is sometimes referred to as constructional stretch. The second phase is a prolonged period of stretch. Only small changes in rope diameter occur, resulting from normal wear and fatigue of the metal under load. The third phase is considered the end-of-life stretch. The rope begins to stretch at a faster pace, much like the first phase. Rapid deterioration of the rope begins as the rope diameter continues to shrink.
- *Reduction in rope diameter* – Stretching, as described above, is one of the primary reasons for a reduction in diameter. However, a reduction can also be caused by excessive external abrasion, internal or external corrosion, or inner wire or core breakage.
- *Peening* – Peening occurs when a wire rope repeatedly strikes another part of the crane, or even repeatedly strikes itself. Another cause is continuous work under high load over a sheave or drum. Both of these causes are usually associated with duty-cycle work such as dredging and pile driving. If the situation cannot be controlled, the rope must be inspected more often and could require more frequent replacement.

To help balance the wear on a rope in the area of sheaves, it is not unusual to shorten the rope somewhat by cutting off the end. The fresh end of the rope is then terminated again as needed, and a fresh section of rope is then running across the sheaves. This prevents running the same section of rope over the same sheaves for too long, forcing a larger section of rope to be removed and disposed of when the wear becomes excessive. Think of it as a planned trimming of the rope, to keep healthy rope running across the sheaves.

A rope can also be reversed end-for-end to balance wear and make the most of the rope. After the active part of a rope has been in operation for a while and shows some wear, reversing the rope places the previously active section on the drum and the fresh portion of the rope becomes the active section.

Figure 19 Example of core protrusion.

2.3.0 Sheave and Load Block Inspection

The sheave and load block are an integral part of the wire rope system. They must be inspected as precisely as wire ropes. Because of the rope's continuous contact with the load block sheaves, these components can have a tremendous effect on the service life of the rope. Integrating the inspection of crane sheaves and the load block into the wire rope inspection program is a prudent and wise approach.

A sheave is designed to support and guide the wire rope. A load block (*Figure 20*) consists of a series of sheaves stacked in a support frame with a hook attached. When examining the sheaves of a load block, check the bearings and their lubrication. Sheaves should turn easily, not wobble, and be lubricated in accordance with the manufacturer's documentation. Sheaves that are not properly lubricated will increase the amount of energy necessary to move a load due to increased friction. Bearings that are worn out from improper maintenance or normal wear can begin to wobble. Wobbling can cause significant vibration of the wire rope, resulting in accelerated wear. The bearing could also fail completely under a normal load.

Other areas of the sheaves to check are the grooves and flanges (*Figure 21*). The flanges must be checked for physical damage that may snag the wire rope or cause it to jump or walk out of the groove when it is rotating. Flaws in the flanges can also damage the rope as it repeatedly passes over the defect, especially under load.

Metal-to-metal contact, such as the contact between the wire rope and the sheave, can be very abrasive. Abrasion can also be increased

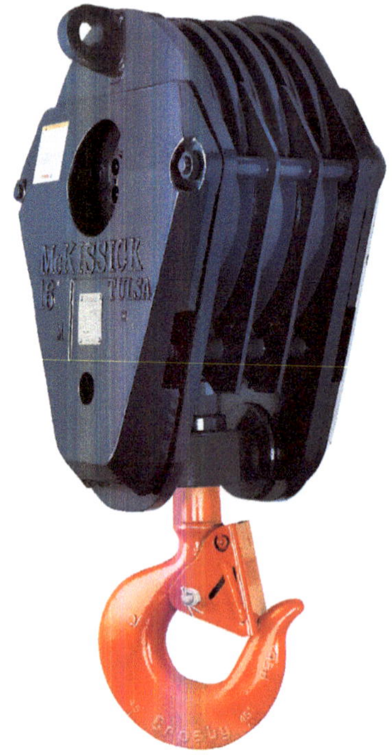

Figure 20 Typical load block.

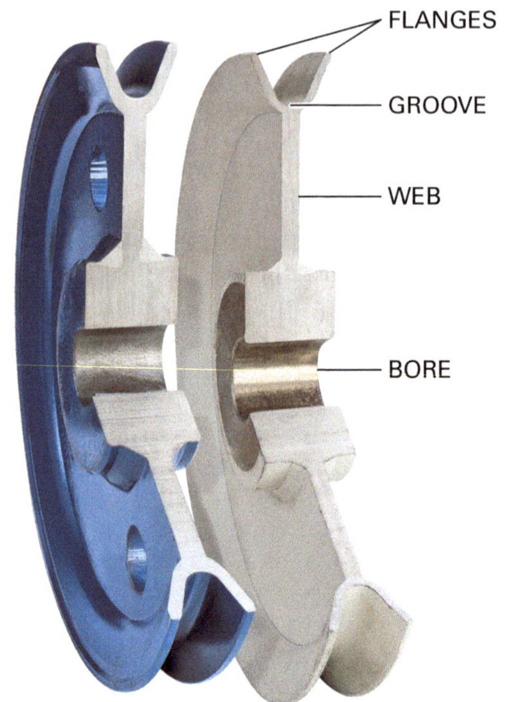

Figure 21 Cross section of a sheave.

significantly if the rope or sheave becomes coated with dirt or sand. It is important that the rope fit the groove properly. If there is only a small area of contact between the two, for example, then the pressure against those areas increases. This can cause increased wear from abrasion as well as distortion of the rope. *Figure 22* shows a wire rope and sheave that are properly matched, where the sheave supports the rope through 120–150 degrees of the rope's circumference.

Regardless of the fit, the wire rope and the sheave will wear each other. As the groove wears, it provides incorrect support for the wire rope, possibly allowing it to flatten out and move deeper into the groove. The range of contact will eventually exceed 150° of the rope's circumference. This increases the speed at which both the wire rope and sheaves deteriorate, since there is too much contact between them.

Sheave grooves are checked for wear using special gauges (*Figure 23*). Each leaf of the gauge is designed to fit a specific size of sheave. There are two types of gauges. A 5 percent gauge is sized to the wire rope plus a 5-percent oversize allowance. There is also a 2.5 percent gauge often used by inspectors. This gauge is considered a go-no go gauge, in that a sheave groove that is smaller than the gauge is rejected for being too tight. Note that worn sheaves can be refurbished by building up the metal and then machining the groove to the size of a new sheave.

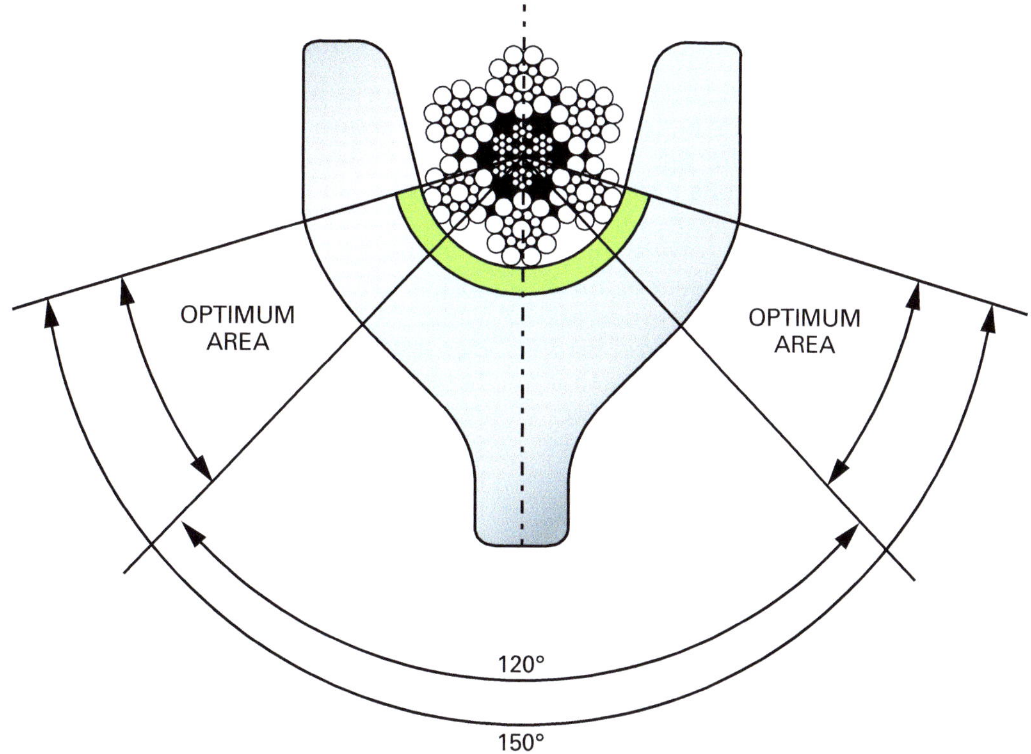

Figure 22 A properly matched sheave provides good support for the rope.

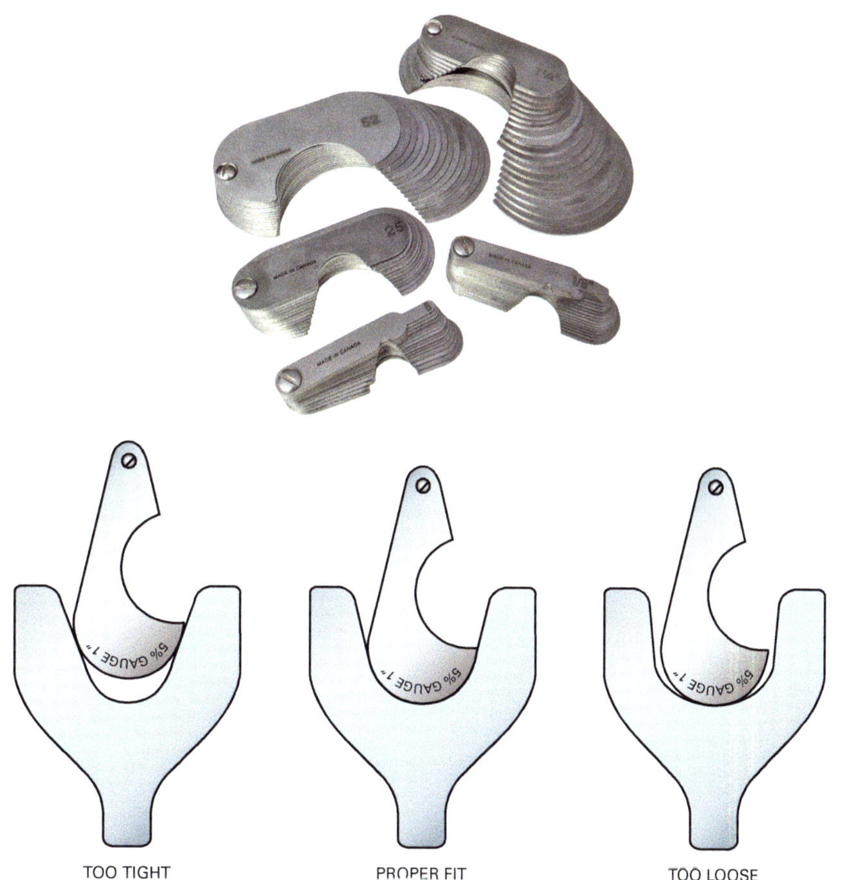

Figure 23 Sheave gauge and use.

2.3.1 Inspecting Hooks

The hook should be examined for damage or flaws, as shown in *Figure 24*. The latch is an essential part of the hook, and tends to be more fragile than other parts of the hook. The latch is spring-loaded so that it closes securely on its own. As it closes, the latch should stop firmly against the tip of a hook. Make sure that the latch is fully intact and not bent or distorted. It should seat against the tip of the hook without a gap, fully closing off the throat. When the hook is attached to a load block with a swivel at the top, check to ensure that the hook bearing is free to rotate in both directions. Double hooks are inspected in the same areas.

2.3.2 Inspecting Load Blocks and Ball Assemblies

Load blocks have sheaves that must be inspected, as well as any reeving guides, side plates, and all assembly hardware. *ASME B30.5, Section 5-1.7.6* requires load blocks to be labeled with their rated capacity and weight. Refer to *Figure 25*

Although a ball assembly may appear to be indestructible, it too must be inspected. Refer to *Figure 26*. Check the ball itself for significant damage, cracks, and excessive corrosion. Latch pins and hook pins should be in good condition with cotter pins installed. The swivel is usually constructed with a heavy-duty bearing. The assembly should rotate smoothly in either direction. Like load blocks, ball and hook assemblies must be clearly labeled with their rated capacity and weight.

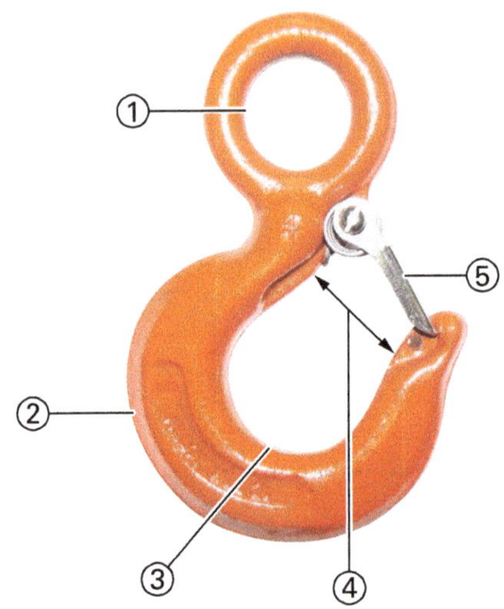

POINT #	HOOK COMPONENT	INSPECTION
1	Eye	Check for cracks, deep nicks, gouges, and any signs of bending or twisting.
2	Body	Check for cracks, deep nicks, gouges, and any signs of bending or twisting.
3	Saddle	Check for cracks, deep nicks, gouges, and any signs of bending or twisting.
4	Throat	Check for stretching or enlarging of the original throat opening.
5	Latch	Ensure the latch is not twisted or otherwise distorted, operates smoothly, and closes off the throat completely.

Figure 24 Inspecting a hook and latch.

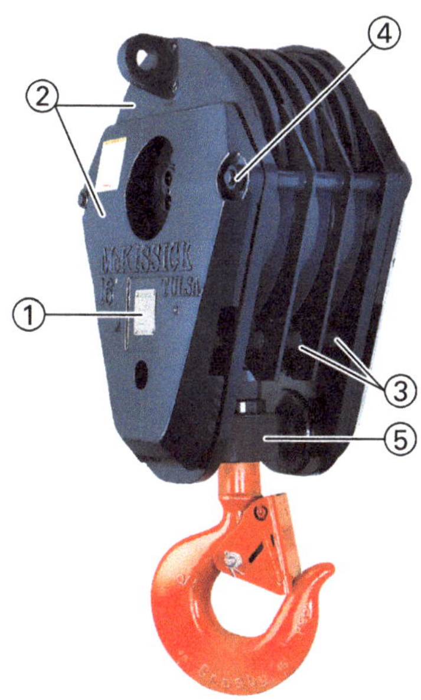

POINT #	HOOK COMPONENT	INSPECTION
1	Data plate	Must be present, legible, and show capacity and weight.
2	Cheek plates/side plates	Check for cracks and noticeable distortion. No severe corrosion or pitting.
3	Reeving guide	Check for bending or other damage. No severe corrosion or pitting.
4	Hardware	Hardware must be tight and secure; no excessive play. No severe corrosion or pitting.
5	Swivel bearing housing	Check for cracks and any sign of bending or twisting. Cotter pins in hook bolt. Ensure that the bearing rotates smoothly in both directions.

Figure 25 Load block inspection.

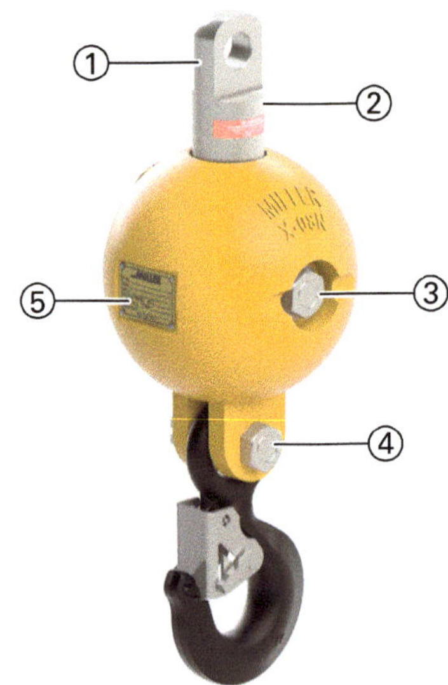

POINT #	HOOK COMPONENT	INSPECTION
1	Swivel eye	Check for cracks, deep nicks, gouges, and any signs of bending or twisting. No severe corrosion or pitting.
2	Swivel	Check for cracks and noticeable distortion. Must be free to rotate in both directions. Check manufacturer's specifications for allowable gap between the swivel and ball assembly. No severe corrosion or pitting.
3	Latch pin	Ensure cotter pins are installed and intact; no excessive play indicating unacceptable wear. No severe corrosion or pitting.
4	Hook pin and bars	Ensure cotter pins are installed and intact; no excessive play indicating unacceptable wear. No severe corrosion or pitting.
5	Data plate	Must be present, legible, and show capacity and weight.

Figure 26 Ball inspection.

2.4.0 Handling and Maintenance Procedures

Information on handling and maintenance practices for wire rope is provided in this section. Remember that wire rope is considered to be a machine in its own right, and this expensive and valuable part of the crane requires the proper care.

2.4.1 Rope Storage

Wire rope is often ordered in anticipation of periodic maintenance, and its replacement is pre-planned. The rope should be stored out of the weather, preferably in a weatherproof area where it will stay dry. Even areas with excessive humidity can accelerate corrosion, but ropes are rarely kept in a humidity-controlled environment. If the area is too hot, or if they are stored near hot

The Wire Rope Technical Board

The Wire Rope Technical Board is an organization comprised of engineers representing the companies that produce over 90 percent of the wire rope manufactured domestically. The group's objectives are to promote the ongoing engineering of wire rope products; assist in the development of standards used in the industry; promote the development of wire rope-related safety standards and their implementation; promote the greater use of wire rope products by providing technical and engineering information; and to conduct and/or support research to the benefit of all users. To that end, the group assembles and distributes its *Wire Rope Technical Manual*, now in its fourth edition.

surfaces, the lubricant may become thin enough to drip out and leave the upper surface dry. The area should be free of corrosive gases and salty air. All wire rope should be stored on a reel until it is installed.

Lubricant can be applied prior to storage for protection. Many ropes are shipped with some type of protective coating already applied. In some cases, serviceable ropes are removed for planned reuse in the future, perhaps on a smaller crane if some portion of it has been removed due to damage. These ropes should be thoroughly cleaned, lubricated and properly stored on a reel. Whether storing the rope on a reel or feeding it on the crane, do not allow the rope to drag through dirt or make contact with anything that might damage it as it moves.

2.4.2 Removing Wire Rope from a Storage Reel

When removing a length of wire rope from a reel, always rotate the reel by supporting it on a reel stand or rolling it along the floor as the rope is removed (*Figure 27*). Avoid removing wire rope from the flanged end of a reel. This creates a kink in the rope for each wrap on the spool. If a kink has formed, no amount of twisting or stress can completely remove it. A kinked rope is weakened and may be unsafe for use. Even a partial kink, called a *dogleg*, will chafe on the flanges of each sheave as it passes over, and could cause premature wear. Manufacturers can provide instructions and guidance for removing rope from a reel, if more information is needed.

Kinks can also form in the wire rope if a section is pulled from the reel and then allowed to go slack. It is best to maintain tension at all times. To do so, the support for the reel needs some type of simple brake that will not allow the reel to turn in either direction on its own. As shown in *Figure 27*, a reel or coil can also be placed on the floor and rolled out in a straight line.

2.4.3 Cutting Wire Rope

There are slightly different rules for cutting standard preformed non-rotation-resistant rope and rotation-resistant rope. Rotation-resistant rope will be discussed at the end of this section. The following applies to non-rotation-resistant rope. Before a rope is cut, seizing is applied to prevent the rope from unraveling and to keep the shape consistent. Seizing the rope before cutting is required by 29 *CFR* 1926.1414 in accordance with the manufacturer's instructions. Standard preformed rope that is non-rotation-resistant has a limited

Figure 27 Unreeling wire rope from spools.

tendency to unravel. For this reason, many require only one seizing to be applied on each side of the cut, although some may require two seizings on each side. To be safe, applying two seizings on each side in all cases is a good idea. Ropes that are 1" in diameter or more may require more seized areas on each side.

Figure 28 shows two methods of wrapping the wire to seize a rope. In the first method, a section of soft wire is wrapped around the rope, with each wrap tightly snugged together. Continue wrapping until the seizing is 1 to $1\frac{1}{2}$ times the diameter of the rope to be cut. For example, if the rope is 1" in diameter, the seized area should be 1" to $1\frac{1}{2}$" long. When placing multiple seizings on each side of the cut area, space them 2 to $2\frac{1}{2}$ times the diameter of the rope apart.

The second method of wrapping begins with both ends of the wire on the same end of the wrap. One end is wrapped around the rope, trapping the other end beneath it. The two ends are then tied together. This keeps the surface of the seizing smooth, without a knot or wire protruding.

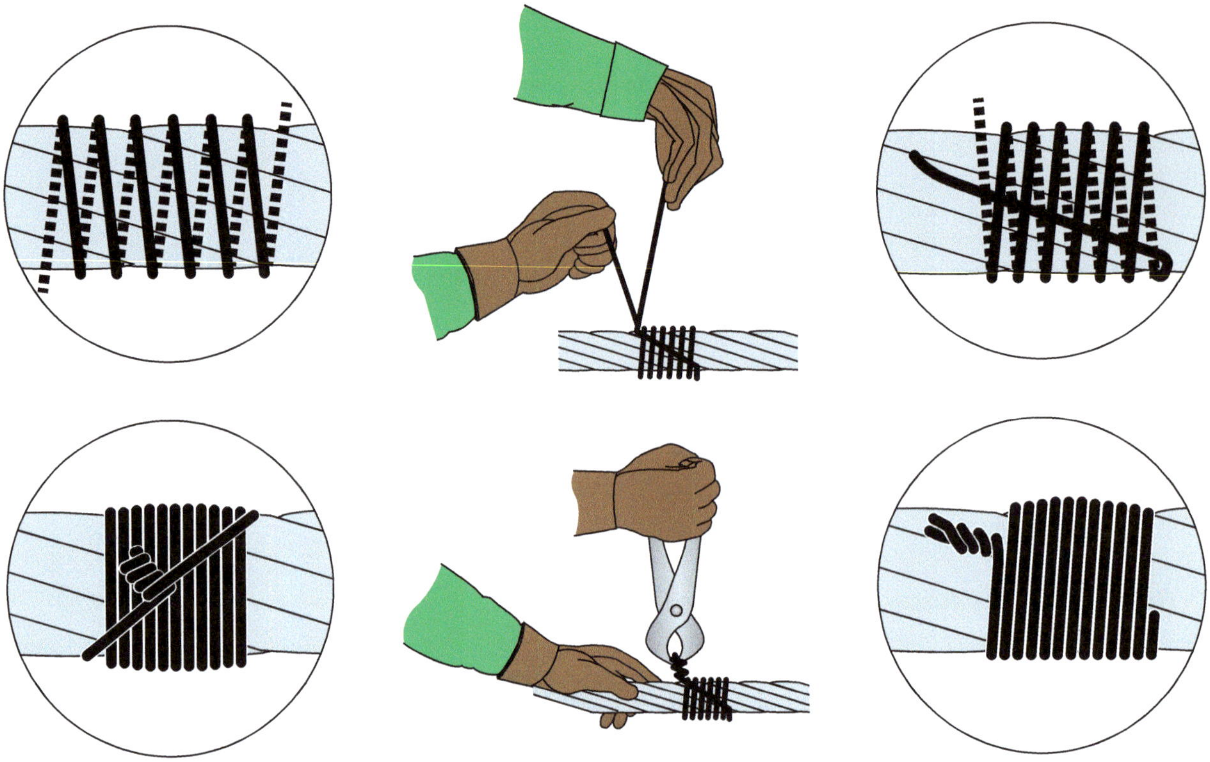

METHOD 1 **TERMINATING WITH A KNOT** **METHOD 2**

Figure 28 Seizing wire rope.

To knot the ends, wrap them around each other a few turns by hand. Use pliers to tighten the twist and remove any slack, then snip the ends and mold or lightly hammer the knot down.

The cut ends of seizing wire are extremely sharp and represent a cut hazard. Be sure that the cut ends and the knot are lightly hammered down. Also note that the cut end of the rope will be very sharp. Handle wire rope cautiously with gloves.

Do not cut wire rope with the edge of a grinding wheel. Grinding wheels are not designed for this task. Use an appropriate abrasive cutting wheel, and ensure that the maximum speed of the wheel is equal to or greater than the speed of the power tool used.

Wire rope is best cut using an abrasive cutting wheel designed for cutting steel. Do not use the edge of a wheel designed for grinding. It may also be cut with an oxyfuel cutting torch, which may result in the rope wires and strands melting and bonding together. If it is cut with a wheel or the end does not fuse soundly when cut with a torch, weld the end of the cable to prevent any possibility of the rope strands loosening and un-raveling (*Figure 29*). The weld diameter cannot be larger than the diameter of the cable. The weld also ensures that the core does not try to move in or out of the rope assembly at the ends. If possible, let the seizing remain on the rope even after the weld is completed.

Figure 29 Wire rope welded after cutting.

Rotation-resistant wire rope requires a bit more care. In many cases, the outer strands are not preformed. This is most common in Category 1 rotation-resistant ropes. As a result, the strands have a greater tendency to unravel. The outer strands must be kept in their proper position with tight seizing to maintain the position of the inner strands and core. Follow the manufacturer's instructions carefully when the rope must be cut. Most manufacturers prefer that rotation-resistant rope not be cut at all, but cuts will generally be needed in order to extend the service life of this expensive component.

For rotation-resistant rope, use a minimum of three seizings on each side of the cut (a total of six seizings). The seizings that are closest to the cut are positioned one rope diameter away from the intended cut. The other two seizings on each side of the cut should be equally spaced at roughly three rope diameters apart.

2.4.4 Lubrication

As wire rope is manufactured, lubricant is applied to the center. The core and the area between it and the strands acts as a storage medium for the oil, which is squeezed out as the rope is used and the wires/strands interact with each other. Proper lubrication applied to the outside of the rope at frequent intervals helps to retain the original lubricant within the rope.

Wire rope tends to collect a lot of dirt and grit in use. Some of the dark grime comes from the degradation of the rope itself and the metal particles that are produced from abrasion. The lubricants used on wire rope can be sticky and they often collect dirt and grit. The rope must be clean before applying lubricant. Wire rope can be cleaned with a wire brush and an appropriate solvent, but it is a challenging task. The liquid solvent must be removed or allowed to evaporate completely. Otherwise, it will quickly break down any new lubricant applied. Steam cleaning is also an option. Once the rope is clean, it should be lubricated immediately to restore corrosion protection.

To lubricate a wire rope, use a lubricant that is light enough to penetrate the outer strands and get to the inner parts of the rope. There are many lubricants on the market for this specific application. Some common lubricants can actually damage the rope, so it is important to follow the recommendations of both the wire rope and lubricant manufacturers. Since the lubricant should be light-bodied, it can be sprayed directly on the rope as it winds on a drum or reel. Remember that OSHA prohibits the use of lubricants that interfere with the ability to visually inspect the rope. A manual pump sprayer, similar to the type used for landscaping purposes, can be used to apply it. Pneumatic spraying systems are best when the task is large and/or continuous. The lubricant can also be brushed on.

It is best to apply the lubricant at a point where the rope is bending on a drum or reel. The flexing of the rope as it bends helps to work fresh lubricant deep into the rope.

2.4.5 Break-In

New wire rope should be broken in after installation. It should be used for about one hour without a significant load to make sure it is compatible with the sheaves and drum. This allows the individual wires and strands to establish their position. After this procedure, rope clips and termination hardware should be checked and retightened if necessary. Remember that constructional stretching will occur for some time. There are wire rope products that are pre-stretched to minimize constructional stretch.

Additional Resources

ASME Standard B30.5, Mobile and Locomotive Cranes. Current edition. New York, NY: American Society of Mechanical Engineers.

ASME Standard B30.9, Slings. Current edition. New York, NY: American Society of Mechanical Engineers.

29 *CFR* 1926, Subpart CC, *Cranes and Derricks in Construction*. Current edition. Washington, DC: US Department of Labor, Occupational Safety and Health Administration.

29 *CFR* 1926.251, *Rigging Equipment for Material Handling*. Current edition. Washington, DC: US Department of Labor, Occupational Safety and Health Administration.

Mobile Crane Safety Manual. 2014. Milwaukee, WI: Association of Equipment Manufacturers.

Willy's Signal Person and Master Rigger Handbook. Current edition. Ted L. Blanton, Sr. Altamonte Springs, FL: NorAm Productions, Inc.

North American Crane Bureau, Inc. website offers resources for products and training, **www.cranesafe.com**.

Union®, a WireCo® WorldGroup brand, **www.unionrope.com**.

2.0.0 Section Review

1. The OSHA standards require that a wire rope (or section thereof) be removed from service if the diameter is reduced in size by ______.

   a. 2 percent
   b. 5 percent
   c. 7 percent
   d. 10 percent

2. Allowing a load to stop suddenly during lifting or lowering, resulting in the load bouncing up and down on the rope, is one cause of ______.

   a. abrasion
   b. kinks
   c. birdcaging
   d. crushing

3. Damaged flanges on a sheave can cause the wire rope to ______.

   a. kink
   b. stretch
   c. birdcage
   d. jump out of the groove

4. Before cutting rotation-resistant wire rope, the rope should be seized at least ______.

   a. six times on each side of the cut
   b. four times on each side of the cut
   c. three times on each side of the cut
   d. once on each side of the cut

3.0.0 WIRE ROPE REEVING

Objective

Explain how to reeve wire rope onto load blocks and drums.

    a. Identify important reeving considerations.
    b. Explain how to reeve wire rope onto a hoist drum.

Performance Tasks

    4. Reeve multiple-part wire rope to a load block.

Trade Terms

D/d ratio: The ratio of a sheave's pitch diameter, represented by D, and the body diameter of the wire rope, represented by d, that is passing over it.

Design factor: The ratio of a wire rope's minimum breaking strength to the maximum load expected to be applied.

Pitch diameter: The diameter of the sheave measured at its root where the bottom of the wire rope makes contact as it travels over. The pitch diameter represents the size of the circle formed by the bottom of the rope if it traveled 360° around the sheave.

One common way to use wire rope on a mobile crane is to reeve it through a load block. Attaching the load block requires threading the wire rope around the sheaves in a specified manner. This process is known as *reeving*. Each span of wire rope passing from one sheave to the next sheave is known as a *part of line*. As more parts of line are added to the load block, the amount of weight pulling on the wire rope at any given point is reduced. This reduction can be attributed to the fact that the total weight of the load will be equally shared between each part of line reeved on the load block.

3.1.0 Reeving Considerations

Before learning how to properly reeve wire rope, there are several factors to consider. It is also helpful to understand the relationship between the wire rope diameter and the sheaves, referred to as the D/d ratio.

3.1.1 D/d Ratio

The D/d ratio is the mathematical relationship between the pitch diameter of the sheave and the rope diameter. This factor is important in establishing the fatigue resistance or relative service life of the wire rope. A small ratio indicates that the wire rope is having to bend sharply to pass over the sheave. When the D/d ratio is too small, it can increase wear significantly on both the rope and the sheave. This wear can be directly attributed to the increased bending stress applied to the rope.

The pitch diameter of a sheave represents the diameter measured at the root of the groove where the rope rides, as shown in *Figure 30*. A simple D/d ratio calculation using 1" wire rope is also shown in the figure. The outer diameter of a sheave includes the sheave flange, which keeps the rope captured in the groove. Only the pitch diameter is applied when determining the D/d ratio; assuming the sheave is designed for wire rope and mobile crane applications, the outer diameter of the sheave is of no consequence.

When the rope diameter is 1", it is quite simple to show the ratio, as shown in *Figure 30*. A bit more effort is required for other rope diameters. To be informative, the ratio must have the number 1 on one side. For example, assume the pitch diameter of a sheave is 16", and a $\frac{3}{4}$" rope diameter is desired. Determine the ratio as follows:

$$D/d \text{ Ratio} = D \div d$$
$$D/d \text{ Ratio} = 16 \div 0.75$$
$$D/d \text{ Ratio} = 21.3, \text{ expressed as } 21.3{:}1, \text{ or } 21.3 \text{ to } 1$$

ASME Standard B30.5 specifies the minimum D/d ratio for sheaves and ropes used for specific applications on the crane, as follows:

- Boom hoist sheaves must have a minimum ratio of 15:1
- Load block sheaves must have a minimum ratio of 16:1
- Load hoist sheaves must have a minimum ratio of 18:1

Note also that the hoist drum must have an 18:1 ratio to the diameter of the rope. Rotation-resistant rope must have an 18:1 ratio in all the cases above.

By comparing the minimum ratio with the calculated ratio, it can be determined if a rope and sheave are compatible with each other in terms of safety and durability. The example above resulted in a ratio of 21.3:1, which is an adequate ratio to satisfy all of the listed applications. Sheave and wire rope manufacturers may also specify minimum ratios for each specific product. If a sheave or rope manufacturer requires a higher ratio than that specified by *ASME Standard B30.5*, then that requirement should be followed. Regardless of the minimum values specified in the ASME standard, do not use a wire rope and sheave combination that results in a ratio less than the manufacturer's minimum.

A calculation of the ratio should be made each time a sheave, drum, or wire rope is replaced with a component other than that recommended by the manufacturer, or with a component having specifications that differ from what was previously installed. The sharp bend in the rope that is forced by using a sheave that is too small (or a rope that is too big) will result in excessive wear and damage to the rope as well as to the sheave. Also, remember that the wire rope must properly conform to the sheave surface.

3.1.2 Determining Parts of Line

Figure 31 shows several reeving patterns and how many parts of line result from the configuration. Note that the dead end of the rope terminates at the load block whenever there is an odd number of parts, and at the boom tip when there is an even number of parts.

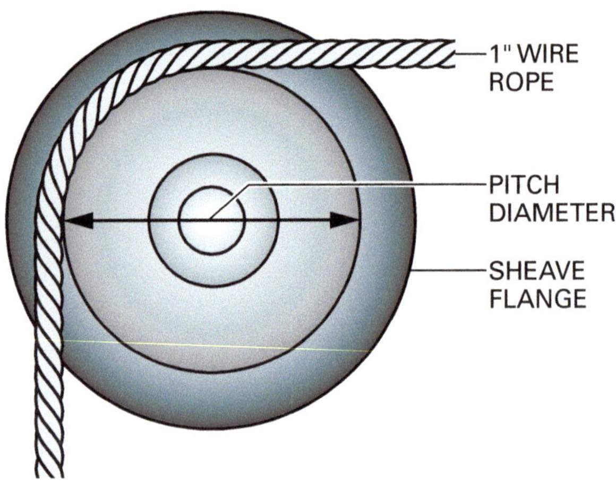

Figure 30 Pitch diameter and the D/d ratio.

Several methods are used to determine how many parts of line are needed to safely accommodate a lift. Some crane manufacturers place this information in the load charts, which is the easiest way to access the needed information. This method is also accurate if all the material used is that recommended by the manufacturer.

Another method used to determine the parts of line needed is a formula and chart provided by equipment suppliers. This method uses a ratio calculated by dividing the total weight of the load to be lifted by the single-line pull capacity of the rope, which is the documented minimum breaking strength of the wire rope. To be safe, the total weight should include everything suspended below the boom tip, including the load block. This ratio is then matched to a table that has the required parts of line listed. Such tables also take into consideration the type of bearing or bushing installed in the sheave in order to account for frictional loads associated with the sheave. Although a rope and sheave combination may be sufficient to perform a lift, the bearing or bushing that allows the sheave to rotate may not meet the required criteria.

Another method used to manually calculate the needed parts of line for a given lift is based on the following formula: minimum breaking strength (as determined by the manufacturer) divided by the design factor equals the allowable load for the wire rope (allowed single-line pull). Written as an equation, the formula looks like this:

$$\text{Minimum Breaking Strength} \div \text{Design Factor}$$
$$= \text{Allowable Load}$$

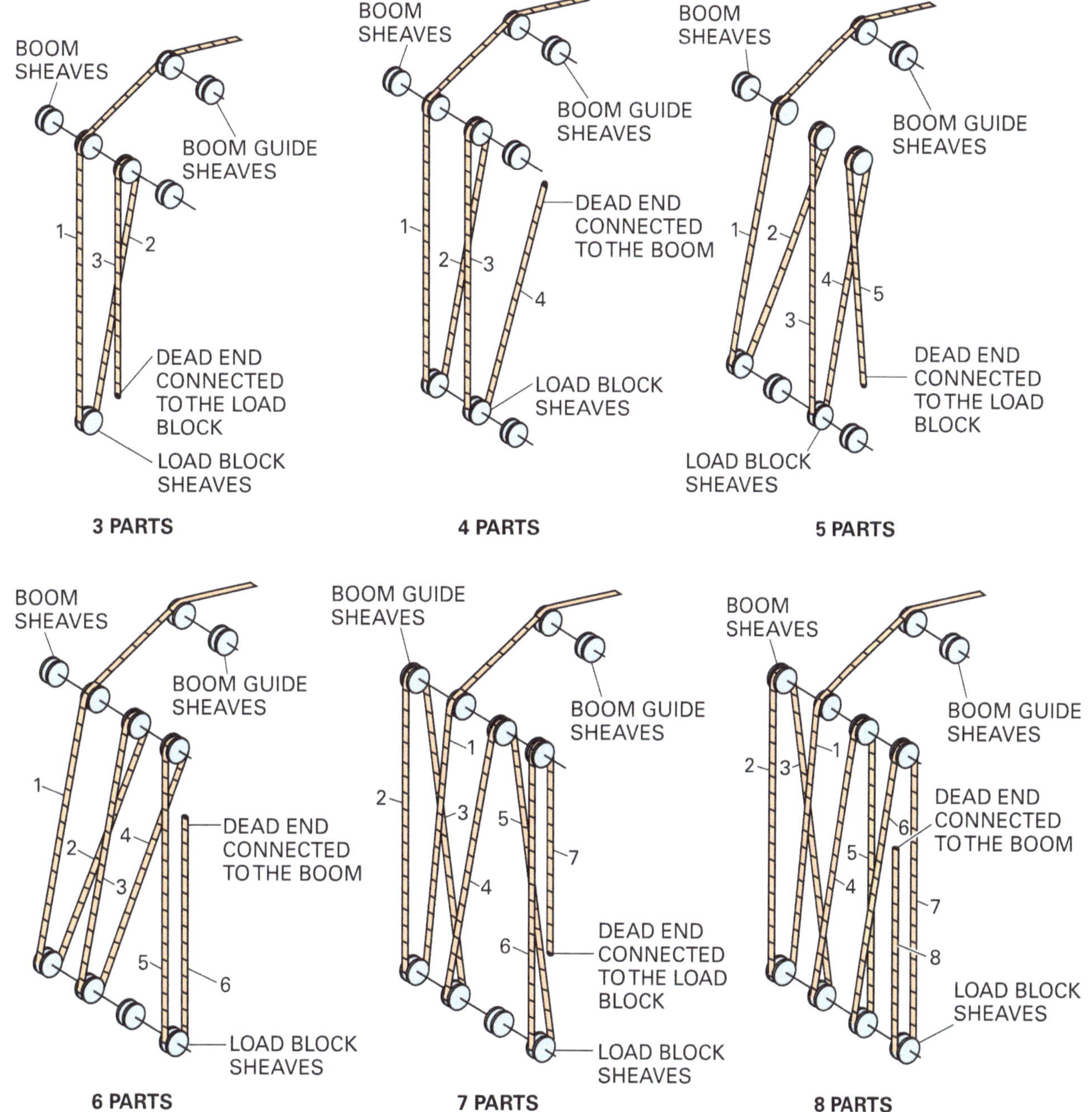

Figure 31 Reeving diagrams.

The design factor expresses the ratio between the minimum breaking strength and the load to be applied. For example, if a wire rope is reported to have a minimum breaking strength of 44,000 pounds (19,960 kg) and the total load below the boom tip is 11,000 pounds (4,990 kg), the design factor (ratio) is 4:1. This is determined by dividing 44,000 by 11,000. *ASME Standard B30.5* requires the following design factors:

- For supporting loads and for supporting the boom and any working attachments at the manufacturer-recommended travel positions and boom lengths:
  - Live or running rope must have a minimum design factor of 3.5.
  - Boom pendants and standing ropes must have a minimum design factor of 3.0.
- For supporting the boom under prescribed boom-erection conditions:
  - Live or running rope must have a minimum design factor of 3.0.
  - Boom pendants and standing ropes must have a minimum design factor of 2.5.
- For rotation-resistant rope, the minimum design factor is 5.0.

3.1.3 Block Twist

Block twist is frequently encountered when using a load block. This problem is directly associated with the construction of wire rope. When a load is placed on the wire rope, it tends to rotate in the opposite direction of the lay, removing the twist inserted during the manufacturing process. As the load increases, the amount of twist also increases. This untwisting can cause a reeved load block to rotate. Rotation of the load block causes severe problems during a lift, up to and including binding between the load block, sheaves, and wire rope.

One way to reduce block twist is by reeving the block to an even number of parts. The dead end of the rope is then terminated at the boom to increase its separation from the other parts of line. Block twist is more prevalent with an odd number of parts, when the rope is terminated at the load block. Using larger sheaves also helps; this reduces the amount of torque applied to the block. A tag line on the load block and/or load can physically prevent twisting.

Rotation-resistant rope is designed to reduce twist and is often used with the manufacturer's approval. A small amount of twist, however, should still be expected.

3.2.0 Reeving a Drum

The component that transfers the lifting power of the crane to the wire rope is the hoist drum. The drum provides the physical connection of the rope to the crane. The hoist drum holds the load, or rotates to move the load up and down as required for the task. Since the interface between the crane and the load is critical to the safe operation of the crane, the method for reeving the wire onto the drum is also critical.

<table><tr><td>**WARNING!**</td><td>Replacement wire rope must have a strength rating at least as high as the original rope furnished or as high as the rope recommended by the manufacturer. Any change from the original size, grade, or construction is to be specified by a rope manufacturer, the crane manufacturer, or an otherwise qualified person. Using a wire rope for hoisting that does not have the proper construction for the application can lead to unexpected rope failures, significant property damage, and serious personal injury.</td></tr></table>

Two types of drums can be encountered on a crane, grooved and smooth. A grooved drum has channels into which the first layer of wire rope is wound. The grooves help ensure that the first layer of rope lays on the drum in a precise position. The second and subsequent layers of wire rope lay in the valleys formed by the coils of the first layer. It is imperative that the drum be reeved correctly. If the wire rope is carelessly wound and jumps from one groove to another, for example, it will be crushed when stress is applied to the other layers on top of it.

A smooth drum has no grooves to guide the first layer. However, just as with the grooved drum, the subsequent layers of the smooth drum lay in the valleys of the previous layer. Correct reeving of the first layer is essential with a smooth drum. Loose and uneven layers result in excessive wear, crushing, and distortion of the rope. This reduces the rope's effective strength, and in some cases, ruins the rope.

Wire rope drums are reeved using one of four directional methods, as shown in *Figure 32*. To determine the proper direction to wind the wire rope, stand behind the crane and point down the wire rope path with the index finger. Use the hand that is in the direction of the lay of the rope; for example, use the left hand for left lay and the right hand for right lay wire rope.

Figure 32 Drum reeving.

If the wire rope is to be wound over the top, referred to as an *overwind*, point with the palm of the hand facing down. If the wire rope is to be underwound, with the rope entering at the bottom of the drum, have the palm of the hand facing up. Using this procedure, the proper direction to apply left and right lay rope on a drum can easily be determined.

Hoist drums must not be overfilled with rope. *ASME Standard B30.5, Section 5-1.3.2(a)(2)(c)*, which is incorporated by reference into the OSHA standard, specifies that the drum flange must extend at least $^1/_2$" (127 mm) beyond the top layer of the rope at all times.

One last critical element for reeving wire rope is its proper removal from the storage reel to the drum. Unwind the rope from the storage reel in the same way that it is being wound onto the drum. Refer to *Figure 33*. For example, if the wire rope is being overwound onto the drum, then the rope should leave the storage reel from the top. If the wire rope is being underwound onto the drum, then the rope should leave the storage reel from the bottom. When underwinding wire rope to a drum from the bottom of a storage reel, it is important to maintain constant tension on the rope. Otherwise, a reverse bend can occur in the wire rope as it becomes slack.

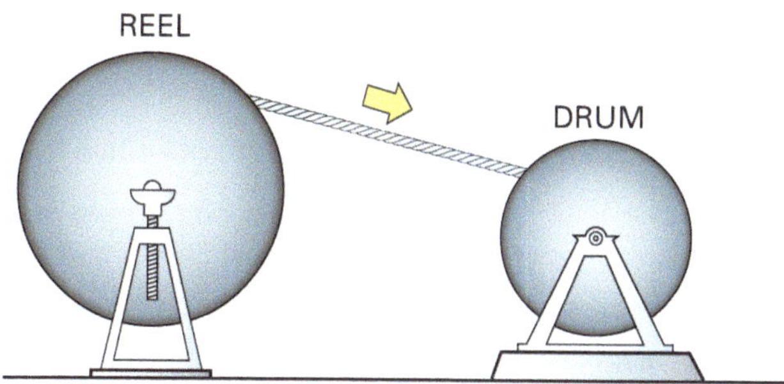

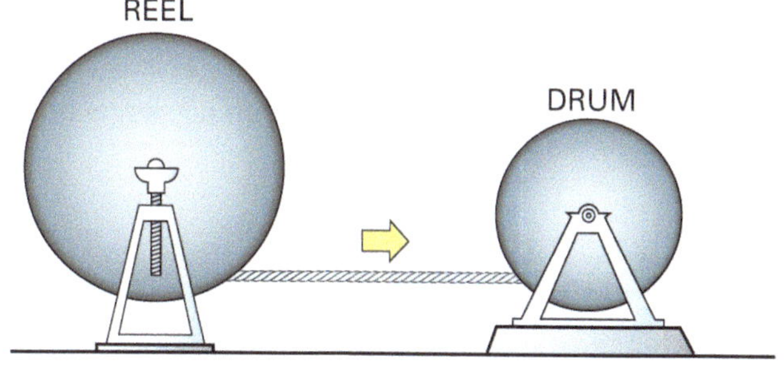

Figure 33 Storage reel-to-drum arrangements.

With the above in mind, follow the basic steps below to install wire rope on a crane:

Step 1   Unload the rope properly, and relieve any twists.

Step 2   Pull the rope over the point sheave and attach the end to the drum.

Step 3   Spool the rope onto the drum slowly and carefully.

Step 4   Spool the first layer tightly. The first layer forms the foundation, or grooving, for the remaining upper layers. If it is not tight, wraps in upper layers will pull down between wraps already on the drum. This can lead to rope damage and reduced rope strength and service life.

Step 5   Spool multiple layers with sufficient tension. The tension load should range from one to two percent of the rope's nominal strength.

Step 6   If the ropes are in a multi-part system, reeve the traveling block and boom tip sheaves so that the rope spacing is maximized. This is done so the traveling block hangs straight and level, and to assure block stability.

Step 7   Check the rope for twists.

Step 8   Cycle the new rope several times, allowing it to form itself to the drum before placing it into service.

> **CAUTION**
>
> As required in 29 *CFR* 1926.1413, there must be a minimum of two wraps remaining on the drum when the load and/or boom is dropped to its lowest point.

Additional Resources

ASME Standard B30.5, Mobile and Locomotive Cranes. Current edition. New York, NY: American Society of Mechanical Engineers.

ASME Standard B30.9, Slings. Current edition. New York, NY: American Society of Mechanical Engineers.

29 *CFR* 1926, Subpart CC, *Cranes and Derricks in Construction.* Current edition. Washington, DC: US Department of Labor, Occupational Safety and Health Administration.

29 *CFR* 1926.251, *Rigging Equipment for Material Handling.* Current edition. Washington, DC: US Department of Labor, Occupational Safety and Health Administration.

Mobile Crane Safety Manual. 2014. Milwaukee, WI: Association of Equipment Manufacturers.

Willy's Signal Person and Master Rigger Handbook. Current edition. Ted L. Blanton, Sr. Altamonte Springs, FL: NorAm Productions, Inc.

North American Crane Bureau, Inc. website offers resources for products and training, **www.cranesafe.com**.

Union®, a WireCo® WorldGroup brand, **www.unionrope.com**.

3.0.0 Section Review

1. The D/d ratio is used to _____.
   a. calculate the length of rope needed
   b. determine whether a sheave size is appropriate for the rope diameter
   c. calculate the number of parts of a line needed
   d. measure the efficiency of a wire rope termination

2. If you are directed to overwind a right lay rope onto a drum, you would _____.
   a. secure the rope on the right side of the drum and wind right to left
   b. secure the rope on the left side of the drum and wind left to right
   c. choose the proper side for attachment based on the D/d ratio of the drum and rope
   d. choose the proper side for attachment based on the rope diameter only

Summary

The critical link between a crane and its load is the wire rope. It allows the power of the crane to be transmitted to the load, resulting in lift and movement. An understanding of wire rope construction, use, and maintenance practices are critical to the safety of the crane, the load, and the members of the lift team. Proper application and handling also allows this expensive asset to achieve its full life expectancy.

During normal operations, the crane must be inspected on a routine basis. The wire rope has its own set of inspection criteria, and it must be inspected and maintained as carefully as any other part of the crane. It will eventually wear out, as most machines do. When replacing wire rope, it is essential that it be the proper type, fits the sheaves well, and is installed in a manner consistent with the wire rope design.

3. The three parts of a wire rope are the wire, the strands, and the ______.

    a. monel
    b. center wire
    c. core
    d. lang

4. Typical wire-rope core types are fiber core (FC), wire strand core (WSC), and ______.

    a. filler core (FFC)
    b. independent wire rope core (IWRC)
    c. rotation resistant core (RR)
    d. Seale core (SC)

5. The four basic strand patterns are wire, Warrington, filler wire, and ______.

    a. independent wire rope
    b. lang
    c. Seale
    d. synthetic

6. What is the key factor in the construction of rotation-resistant wire rope?

    a. Rotation-resistant rope is constructed with far fewer strands than any other type.
    b. All rotation-resistant rope is constructed using preformed wire and preformed strands.
    c. The lay of the inner strands is precisely the same as the outer strands.
    d. The lay of the inner strands is opposite of the lay of the outer strands.

7. Which wire rope below is made from the weaker type of steel, based on the following label information?

    a. 1,100 feet $\frac{3}{4}$" 6 × 19 F pref RRL EEIP IWRC
    b. 1,100 feet $\frac{1}{2}$" 6 × 25 WS pref RLL IPS FC
    c. 1,100 feet $\frac{1}{4}$" 6 × 26 S pref LLL EIPS IWRC
    d. 1,100 feet $\frac{5}{8}$" 6 × 37 WS pref RRL PS IWRC

8. Which of the following rope terminations have an efficiency rating of less than 100 percent?

    a. Poured socket
    b. Swaged socket
    c. Thimble with wire rope clips
    d. Thimble with mechanical splice

9. A fabricated socket (poured or swaged) has an efficiency rating of ______.

    a. 70 to 95 percent
    b. 75 to 80 percent
    c. 80 percent
    d. 100 percent

10. When assembling a wedge socket on running rope, the tail length of the dead end should be at least ______.

    a. long enough to accommodate three rope clips
    b. three rope diameters but not less than 3"
    c. six rope diameters but not less than 6"
    d. as long as the complete wedge-socket assembly

11. Which OSHA-defined category of wire-rope deficiencies is considered the most significant, meaning that the rope must either be replaced or the damaged section cut off, regardless of manufacturer input?

    a. Category I
    b. Category II
    c. Category III
    d. Category IV

12. A break in a single wire that, when closely examined, has a tapered appearance with one tip cup-shaped (concave) while the other tip is cone-shaped (convex) is likely the result of ______.

    a. loads that exceeded its tensile strength, *i.e.*, stretched to failure
    b. crushing while on the drum in a lower layer
    c. simple fatigue from normal wear and tear
    d. abrasion against other objects

13. A proper fit between a wire rope and a sheave results in the sheave supporting the wire rope through ______.

    a. 60 to 90 degrees of the rope's circumference
    b. 90 to 120 degrees of the rope's circumference
    c. 120 to 150 degrees of the rope's circumference
    d. 135 to 180 degrees of the rope's circumference

14. The mathematical relationship between the pitch diameter of the sheave and the rope diameter is referred to as the ______.

    a. datum
    b. D/d ratio
    c. sheave function
    d. bend ratio

15. The easiest method for determining the number of parts of line needed for a lift is to ______.

    a. check the crane load chart
    b. multiply the load weight by the rope breaking strength
    c. check the OSHA standards
    d. calculate the D/d ratio

16. Which of the factors below increases a non-rotation-resistant rope's tendency to twist in the opposite direction of the lay?

    a. Increasing the load on the rope
    b. Larger rope diameter
    c. A large D/d ratio
    d. Type of rope lubricant used

17. *ASME Standard B30.5* prohibits overfilling of a drum with wire rope by specifying that the drum flange must extend past the top layer of rope by ______.

    a. 2" (508 mm)
    b. 1" (254 mm)
    c. $\frac{1}{2}$" (127 mm)
    d. $\frac{1}{4}$" (64 mm)

Birdcaging: A term used to describe the condition where wire rope is forced into compression along its length, pushing the outer strands away from the core and/or inner strands, forming a shape that resembles a bird cage.

Constructional stretch: The stretching that occurs in wire rope when it is initially placed in service, during its break-in period. Constructional stretch is expected in all wire ropes. Constructional stretch in standard ropes is generally 0.25 to 1 percent of the rope's length.

Core: The axial part at the center of wire rope around which the individual strands are laid.

Crossover points: Per 29 *CFR* 1926.1401, specific locations where a wire rope layer spooled on a drum must climb up and cross over the previous layer. Crossover points are located at each end of the drum at its flanges, where the rope must change direction to continue spooling onto the drum.

D/d ratio: The ratio of a sheave's pitch diameter, represented by D, and the body diameter of the wire rope, represented by d, that is passing over it.

Dead end: The excess wire rope exposed after a termination has been properly fitted. This part of the wire rope is not exposed to the strain applied by a load to the wire rope.

Design factor: The ratio of a wire rope's minimum breaking strength to the maximum load expected to be applied.

Fatigue fractures: Progressive fractures resulting from the repeated bending of individual wires. These fractures may occur at bending-stress levels well below the ultimate strength of the material. Lack of lubrication and rust on a wire rope can accelerate fatigue fractures.

Fiber core (FC): A cord or rope of natural or synthetic fiber used as the axial center (core) of a wire rope.

Filler wire: Small wire used to create spacers within a strand that help hold the position and support the other strand wires.

Flange points: Per 29 *CFR* 1926.1401, points where the wire rope contacts the drum flanges at each end.

Independent wire rope core (IWRC): A wire rope used as the axial center (core) of a larger wire rope assembly.

Lay: In the context of wire rope, the lay describes the direction of rotation of the strands in a wire rope assembly.

Lay length: The distance measured parallel to the axis of the rope or strand in which a strand or wire makes one complete revolution around the core or center.

Live section: Any part of a wire rope that is exposed to the load weight.

Minimum breaking strength: Also referred to as *minimum breaking force*. The documented breaking strength of a wire rope product established under set industry testing procedures.

Monel™: A trademarked metal alloy made primarily from nickel and copper, with small amounts of other materials added. Monel is resistant to corrosion initiated by a number of common sources, including seawater, and is an expensive material.

Pendants: Fixed lengths of rope with mechanical fittings at each end.

Pitch diameter: The diameter of the sheave measured at its root where the bottom of the wire rope makes contact as it travels over. The pitch diameter represents the size of the circle formed by the bottom of the rope if it traveled 360° around the sheave.

Preformed strand: A strand constructed of wires that were formed into the helical shape they would assume once placed in the strand, before the strand was assembled.

Preformed wire rope: A wire rope constructed of strands that were formed into the helical shape they would assume once placed in the rope, before the rope was assembled.

Repetitive pickup points: Per 29 *CFR* 1926.1401, refers to the sections of a rope where short-cycle operations cause it to be repeatedly spooled on and off a small portion of the drum.

Rotation-resistant wire rope: Wire rope constructed with inner and outer strands placed in opposing lays, rather than using the same lay in both layers. This causes the torque of the strand layers to oppose each other, significantly reducing the tendency to twist.

Running ropes: Wire ropes that move over sheaves and/or drums.

Seizing: The securing of an open end of wire rope by wrapping wire tightly around its circumference several times. This action prevents fraying and unraveling of the wire rope. This preparation is used on both sides of a planned wire rope cut.

Speltered: To be attached with molten zinc or a zinc alloy. The term spelter is considered a synonym for zinc.

Strands: A single grouping of round or shaped wires laid helically around a core or center wire. A wire rope assembly for lifting purposes is typically a collection of strands around a common core.

Strand patterns: The various configurations of the wires used to form a strand. The four basic strand patterns are wire, Seale, Warrington, and filler wire.

Swaged: A connection made by forming metal, usually with a die or shaped tool, to change the diameter. A swaged wire rope connection is made by reducing the diameter of the socket around the rope.

Wire: A single, continuous length of metal, usually cold-drawn or cold-rolled from a rod, available in various diameters.

Wire strand core (WSC): A wire strand assembly used as the axial center (core) member of a wire rope.

Additional Resources

ASME Standard B30.5, Mobile and Locomotive Cranes. Current edition. New York, NY: American Society of Mechanical Engineers.

ASME Standard B30.9, Slings. Current edition. New York, NY: American Society of Mechanical Engineers.

29 *CFR* 1926, Subpart CC, *Cranes and Derricks in Construction*. Current edition. Washington, DC: US Department of Labor, Occupational Safety and Health Administration.

29 *CFR* 1926.251, *Rigging Equipment for Material Handling*. Current edition. Washington, DC: US Department of Labor, Occupational Safety and Health Administration.

Mobile Crane Safety Manual. 2014. Milwaukee, WI: Association of Equipment Manufacturers.

Willy's Signal Person and Master Rigger Handbook. Current edition. Ted L. Blanton, Sr. Altamonte Springs, FL: NorAm Productions, Inc.

North American Crane Bureau, Inc. website offers resources for products and training, **www.cranesafe.com**

Union®, a WireCo® WorldGroup brand. **www.unionrope.com**

Figure Credits

Link-Belt Construction Equipment Company, Module opener

The Crosby Group LLC, Figures 13, 14, 20, 21, 25 (photo)

Columbus McKinnon Corporation, Figures 15, 24 (photo)

Lift-All Company, Inc., Figure 17

© Unirope® - Canada, Figure 19

WR Sheave Gauges Ltd., Figure 23 (photo)

Miller Lifting Products of Charlton, MA USA, Figure 26 (photo)

Answer	Section Reference	Objective
Section One		
1. a	1.1.1	1a
2. c	1.2.0	1b
Section Two		
1. b	2.1.3	2a
2. c	2.2.0	2b
3. d	2.3.0	2c
4. c	2.4.3	2d
Section Three		
1. b	3.1.1	3a
2. b	3.2.0; Figure 30	3b

NCCER CURRICULA — USER UPDATE

NCCER makes every effort to keep its textbooks up-to-date and free of technical errors. We appreciate your help in this process. If you find an error, a typographical mistake, or an inaccuracy in NCCER's curricula, please fill out this form (or a photocopy), or complete the online form at **www.nccer.org/olf**. Be sure to include the exact module ID number, page number, a detailed description, and your recommended correction. Your input will be brought to the attention of the Authoring Team. Thank you for your assistance.

Instructors – If you have an idea for improving this textbook, or have found that additional materials were necessary to teach this module effectively, please let us know so that we may present your suggestions to the Authoring Team.

NCCER Product Development and Revision

13614 Progress Blvd., Alachua, FL 32615

Email: curriculum@nccer.org
Online: www.nccer.org/olf

❑ Trainee Guide ❑ Lesson Plans ❑ Exam ❑ PowerPoints Other ___________________

Craft / Level: _______________________________________ Copyright Date: _____________

Module ID Number / Title: ____

Section Number(s): ____

Description: ____

____

____

____

Recommended Correction: ____

____

____

____

Your Name: ____

Address: ____

____

Email: _______________________________________ Phone: _____________________

Telescopic-Boom Attachment Setup and Assembly

OVERVIEW

Manufacturers build and sell cranes in various models with specific options. However, operators can greatly increase the utility of cranes by installing add-on components provided by manufacturers. This module describes the assembly and disassembly of several types of telescopic-boom extensions and accessories.

Module 21302

Telescopic-Boom Attachment Setup and Assembly

Objective

When you have completed this module, you will be able to do the following:

1. Describe the process of setting up, assembling, and disassembling telescopic-boom components.
   a. Explain how to set up and stow a swing-away lattice extension.
   b. Explain how to set up an auxiliary-sheave boom head.
   c. Identify pre- and post-assembly considerations.
   d. Explain how to install and remove intermediate boom sections.
   e. Discuss the design and operations of manual and power luffing jibs.

Performance Tasks

Under the supervision of your instructor, you should be able to do the following:

1. Analyze the site for adequate assembly/disassembly space.
2. Analyze and determine counterweight requirements needed for boom assembly.
3. Determine the ground stability and ground support needed for assembly and disassembly.

Trade Terms

A/D director	Fly	Manual luffing jib
Boom-extend mode	Level luffing	Power luffing jib
Connector pin	Luffing	Stop rail

Industry Recognized Credentials

If you are training through an NCCER-accredited sponsor, you may be eligible for credentials from NCCER's Registry. The ID number for this module is 21302. Note that this module may have been used in other NCCER curricula and may apply to other level completions. Contact NCCER's Registry at 888.622.3720 or go to **www.nccer.org** for more information.

Contents

Figures

1.0.0 TELESCOPIC-BOOM ATTACHMENT SETUP AND ASSEMBLY

Objective

Describe the process of assembling and disassembling telescopic-boom components.

a. Explain how to set up and stow a swing-away lattice extension.
b. Explain how to set up an auxiliary-sheave boom head.
c. Identify pre- and post-assembly considerations.
d. Explain how to install and remove intermediate boom sections.
e. Discuss the design and operations of manual and power luffing jibs.

Performance Tasks

1. Analyze the site for assembly/disassembly room.
2. Analyze and determine counterweight requirements needed for boom assembly.
3. Determine the ground stability and ground support needed for assembly and disassembly.

Trade Terms

A/D director: The individual specified by OSHA standards to be responsible for the assembly and disassembly of cranes.

Boom-extend mode: Refers to the order, and percentage, of extension telescopic-boom sections. The standard mode is to extend all segments the same percentage. Other modes may selectively extend certain sections to provide different capacities for a given load radius.

Connector pin: Any solid, cylindrical, metal part used to fasten, support, or attach the various components and members of a crane's boom and other hoisting gear.

Fly: With reference to mobile cranes, a boom extension carried on the side of the main boom when not in use, that can swing into position to extend the main boom; also called a swing-away.

Level luffing: The process of luffing the boom of a crane while adjusting the hoist cable and boom extension offset so that the load travels in a horizontal direction.

Luffing: Raising or lowering a crane's boom. A luffing jib can luff independently of the main boom.

Manual luffing jib: A semi-fixed jib extension whose offset must be manually set between lifts.

Power luffing jib: A jib extension that has the means to change its offset during a lift by remote control of the operator.

Stop rail: A heavy metal guiderail used by some crane manufacturers to help align a swing-away to its transport location as it is being stowed.

Many mobile crane accidents result from human errors made during crane assembly. For this reason, operators and owners must make sure workers set up, assemble, and disassemble boom attachments properly. Specific guidelines for installing various attachments are included in the manufacturer's manual for each telescopic-boom crane. This module presents considerations for setting up and taking down jib extensions, and describes the assembly and disassembly of intermediate sections and powered jibs.

Deploying a simple folding jib is considered a setup activity rather than an assembly/disassembly operation. More challenging tasks, such as assembling intermediate sections of a significant boom extension, do fall into the latter category. In such cases, OSHA requires that an assembly/disassembly supervisor, referred to as the A/D director, be involved to ensure proper and safe assembly.

This module begins by presenting information on component deployments that are classified as setup tasks. More complex tasks that require an A/D director are discussed later in the module.

1.1.0 Swing-Away Lattice Extension

A swing-away lattice extension, sometimes called a *fly* (or *swing-away*) is a jib extension that stows in brackets on the base section of the telescopic boom. The extension and its parts are built for each model crane and are serial-number specific. The operator should only operate the crane using those parts of the extension with the same serial number or other identifier as the crane. If the crane is equipped with a load moment indicator (LMI), operators must program it for this specific extension only. If any other extension is used, the LMI cannot function properly and damage may occur.

Figure 1 shows a stowed jib extension, while *Figure 2* shows a two-piece adjustable fly in use. Operators can assemble a jib as a straight extension of the boom or they can set it at various offsets to the boom as a luffing jib.

29 *CFR* 1926.1403–1406 outline safety requirements that apply when assembling or disassembling a crane. However, OSHA's interpretation of the standards indicates that preparing a swing-away jib is not an assembly process, but is instead part of setting up an assembled crane. Therefore, an A/D director is not required for this task.

1.1.1 Installing a Swing-Away Lattice Extension

The length of nonadjustable swing-away extensions can vary depending on the type of crane and the manufacturer. Some older cranes may be equipped with both a swing-away and an A-frame jib on the side of their main booms, but this is much less common today. *Figure 1* is an example of a rough-terrain crane with only a swing-away extension.

Figure 1 Stowed jib extension.

Figure 2 Two-piece adjustable fly assembled at an offset.

Operators must always follow the manufacturer's setup instructions and procedures for their crane. Note that the procedure described here applies to a specific crane model only. Other cranes may require different steps, but overall, the process will be similar. The following is the general procedure for installing a nonadjustable swing-away extension on a telescoping-boom crane.

The operator and any other personnel involved in boom assembly should observe the proper warning and safety information provided in the crane manual or site safety manual.

Steps 1 through 5 prepare the crane to set up the swing-away extension:

Step 1　Extend and set the outriggers or crawlers according to the manufacturer's recommendations. Install proper counterweights as necessary for the intended load and boom length.

Step 2　Fully retract all boom sections, and lower the boom to minimum elevation to permit access to the boom nose.

Step 3　Rig either the main hoist or optional auxiliary hoist cable for a single part of line. Lay the cable off to the side opposite of the boom from the swing-away (normally the left side) while erecting the swing-away.

Step 4 To deploy the stop rail of the swing-away support bracket, remove the retaining clip from the stop rail connector pin, and remove the pin (*Figure 3*). Hold the free end of the stop rail with your hand to prevent the stop rail from swinging open.

Connector pins used on cranes come equipped with some form of retainer to prevent them from falling out. These devices are typically a reusable spring-wire retaining clip inserted into a hole drilled through the pin, or a cotter key that is disposed of after each use. Operators should replace missing retainers immediately. Never use connector pins without the proper retainer in place.

Step 5 Rotate the stop rail until it is in a horizontal position, and insert the stop rail connector pin into the outer pin hole (*Figure 4*). Secure the stop rail pin in place with its retaining clip. Check that the section is secure with the stop rail pin and a pivot pin in the bearing point on the stop rail.

Steps 6 through 9 separate the swing-away from the A-frame jib (if present). These steps also prepare the swing-away for setup on the boom nose. The part numbers referenced are shown in *Figure 5*.

Step 6 Ensure that the A-frame is properly secured. At the front support bracket (6), check engagement of the lock pin (1) and its lock plate (3); at the rear bracket (7), check the lock pin (2) and its lock plate (4).

Step 7 Detach the swing-away from the A-frame jib. At the rear end of the A-frame (5), remove the pin (8), which secures the swing-away (9) to the A-frame at that end. Stow the pin in the A-frame attaching link (13).

Step 8 At the front end of the A-frame (5), remove the pin (10), which secures the swing-away to the A-frame at that end. Lower the attaching link (11) to its stowed-position bracket (12) on the A-frame, and secure it with the pin (10).

Step 9 Unlock the swing-away. At the swing-away rear support bracket (7), pull the lock pin (15), and swing the latch plate (16) out of the way.

Figure 3 Stop rail viewed from under stowed extension.

Figure 4 Extension-boom stop rail in the raised position.

Steps 10 through 15 position the swing-away on the boom nose; refer to *Figure 6*.

Step 10 With the front support bracket (6) acting as a pivot, pull outward on the swing-away tip to rotate the swing-away anchor points (18) onto the attach fittings (17) on the right side of the boom nose.

Step 11 Remove the retaining clips and pull the two attach pins (14) from their stowage brackets in the base of the swing-away. Insert the pins through the attach and anchor fittings on the right side of the boom nose. Reinstall the retaining clips in the pins.

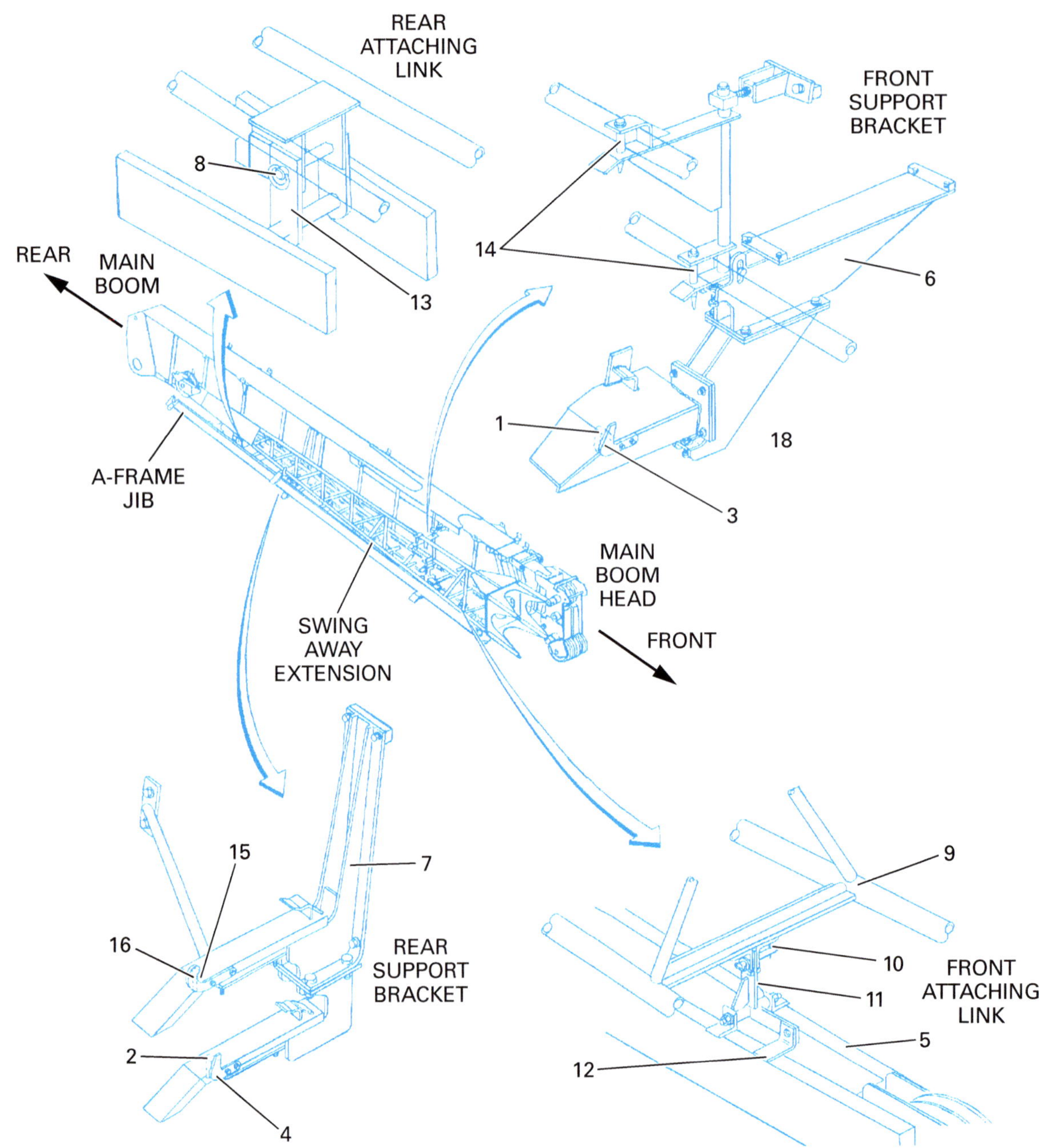

Figure 5 Main boom, swing-away, and support brackets.

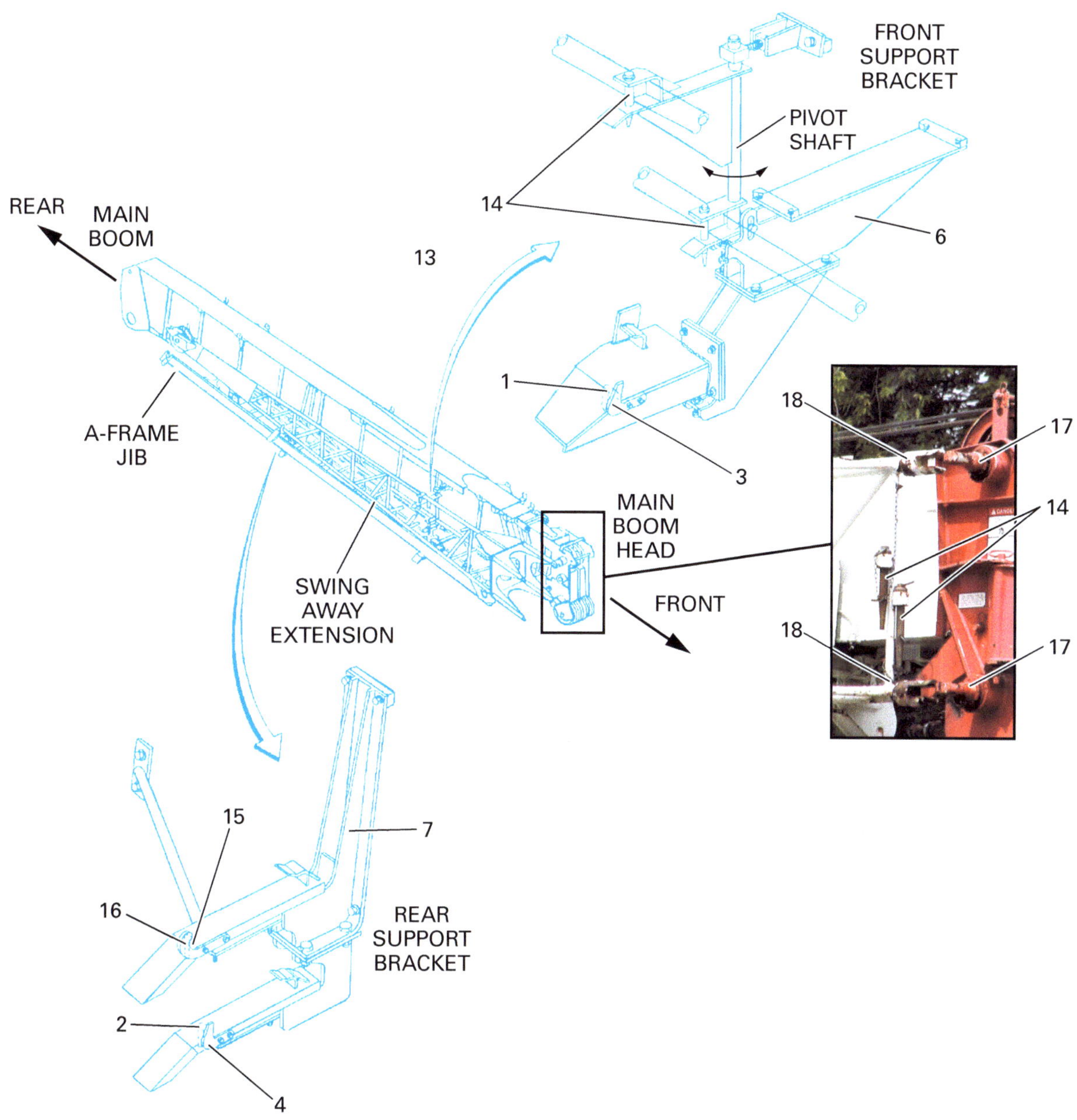

Figure 6 Positioning the swing-away.

Step 12 Attach a tag line to the swing-away. This ensures workers can safely control the motion of the swing-away extension while positioning it.

Step 13 Raise the boom to a level, horizontal position. Then, remove the two connector pins (14) and their retaining clips from the front support bracket (6).

Step 14 Rotate the swing-away into place in front of the boom nose, engaging the anchor points (18) with the attach fittings (17) on the left side of the boom nose (*Figure 7*).

Step 15 Lock the swing-away to the main boom by inserting the two connector pins (14) into the anchor (18) and attach fittings (17) on the left side of the boom. Reinstall the retaining clips in the pins.

Step 16 Remove a pin (19) and retaining clip from the cable-roller sheave (20). Move the cable-roller sheave (20) into working position in accordance with the prescribed inclination. Lock it into place with a pin (19), and secure the pin with its retaining clip.

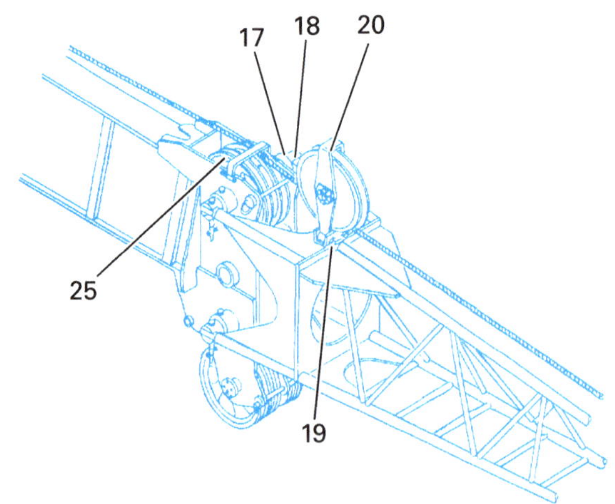

Figure 7 Securing the swing-away on the boom nose.

Step 17 Reel off the hoist rope from the main or auxiliary hoist.

Step 18 Lay the hoist cable over the boom-head sheave (25), under the cable-roller sheave (20), and over the extension tip sheave (24). Remove the cable retaining pins (21) first.

Step 19 Remove the pin (22) and retaining clip securing the cable guide (23) in the stowed position. Position the cable guide in the vertical position, with the cable passing through the guide. Reinstall the pin through the second hole in the guide, and reinstall the retaining clip.

Step 20 Reinstall the cable-retaining pins (21).

Step 21 Reeve the hoist cable on the headache ball.

Step 22 Make the electrical connections to the main boom.

Step 23 Install the anemometer (if equipped) at the head of the swing-away, and electrically connect it.

Step 24 Attach the anti-two-block switch weight. Feed the hoist rope through the weight.

Step 25 Stow the stop rail in the transport position by reversing Steps 4 and 5.

NCCER – *Intermediate Rigger*

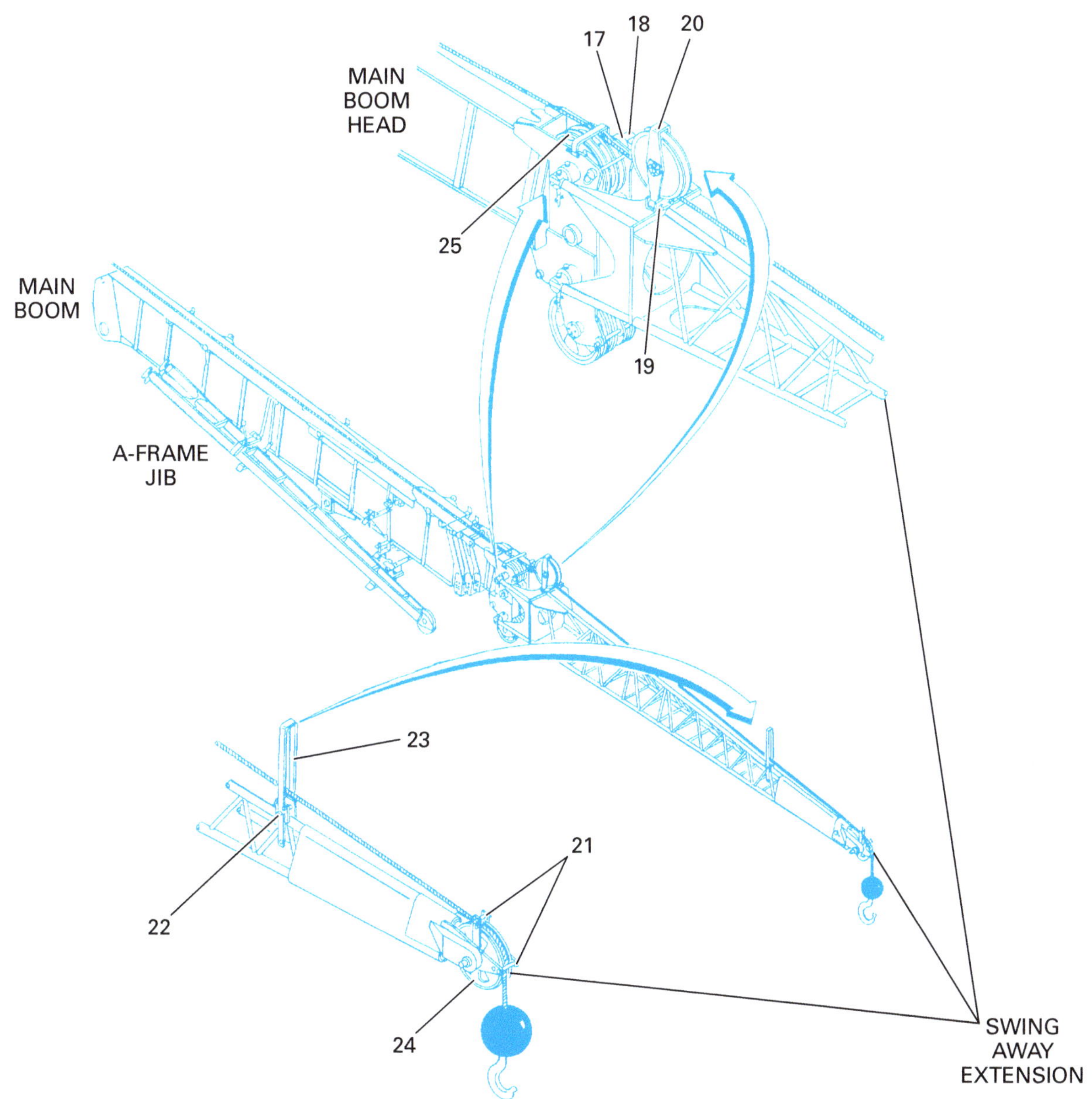

Figure 8 Reeving the hoist line to the swing-away.

1.1.2 *Stowing a Swing-Away Lattice Extension*

Always follow the manufacturer's instructions for stowing a swing-away lattice extension. For general guidelines, refer to the following abbreviated procedure:

Step 1 Fully retract the telescopic sections of the boom.

Step 2 Turn the superstructure to the back or the side.

Step 3 Lower the main boom until the load moment indicator (LMI) switches off.

Step 4 Override the LMI.

Step 5 Lower the main boom until the head sheave is approximately five feet above the ground.

Step 6 Unreeve the hoist rope from the load block, and remove it from the head sheave.

Step 7 Remove the lifting limit switch weight.

Step 8 Electrically disconnect the anemometer (if equipped), then remove it.

Step 9 Move the main boom into 0-degree position, if necessary.

Step 10 Disconnect the extension's electrical connections to the main boom.

Step 11 Move the extension's cable-roller sheave into the transport position.

Step 12 Wind in the hoist rope.

Step 13 Move the stop rail into the assembly position.

Step 14 Attach a tag line to the swing-away.

Step 15 Remove the connector pins between the swing-away and main boom on the left-hand side of the boom head. Place them in their storage brackets on the extension and reinsert their retaining clips.

Step 16 Rotate the swing-away on the main boom until the pivot points on the pivot bearing align.

Step 17 Remove the pins between the swing-away and main boom on the right-hand side of the boom head.

Step 18 Insert the pins in the boom clips at the pivot bearing and reinsert their retaining clips.

Step 19 Rotate the swing-away to the transport position on the pivot bearing. At the rear support bracket, install the latch lock pin to secure the head of the swing-away.

Step 20 Stow the stop rail in transport position.

1.2.0 Auxiliary Boom Head

Crane booms may require an additional piece of equipment called the auxiliary boom head (*Figure 9*). Auxiliary hoists allow lifting relatively light loads without the need to use the main hoist block, which involves extra wear and tear on the main hoist system. Auxiliary boom heads on modern cranes may be one-, two-, and three-sheave models. To attach the auxiliary head to the boom, configure the crane as follows:

- On outriggers or crawlers
- Superstructure swung to a suitable position
- Main boom retracted and set horizontal

Setup requires the same preparations as those for installing boom extensions. The following procedure is a general summary of the steps required. Since these components are quite variable in shape and how they are attached, operators must refer to their crane's manual to install auxiliary boom heads.

Figure 9 Auxiliary boom head with a hoist block.

Step 1 Install the auxiliary boom head in position. Insert the required connector pins and secure them with their retaining clips.

- This step may involve swinging up the component from its transport position below the main boom head to in front of it. In other cranes, you may have to unpin the auxiliary head on one side of the main boom and rotate it so that it is in front of the main boom sheaves. Some cranes do not transport their auxiliary boom head at all. In those cases, an auxiliary crane must lift the component into position to install it.

Step 2 Install the auxiliary head's signal cable(s) into the appropriate socket(s) on the main boom head.

Step 3 Remove the hoist rope guide bars from the head of the main boom and the auxiliary boom head.

Step 4 Lay the hoist rope over the auxiliary boom head and over the appropriate main boom sheave(s).

Step 5 Reinsert the rope guide bars, and secure them with their retaining pins.

Step 6 Install an anti-two-blocking weight on the auxiliary boom head (and on the lifting unit switch on the main boom, if not already installed).

Step 7 Guide the hoist rope through the anti-two-blocking weight. Note that, depending on the anti-two-blocking weight configuration, operators may be able to complete Step 7 after attaching the headache ball.

Step 8 Attach the headache ball on the auxiliary hoist rope.

Step 9 For attaching hook blocks, reeve the auxiliary hoist rope through the hook sheave and attach it to the auxiliary boom head per manufacturer's instructions.

1.3.0 Pre- and Post-Assembly Considerations

While the previous procedures are considered setup, the installation of intermediate jib sections and luffing jibs is considered assembly and disassembly. These tasks have additional requirements.

Company or project supervision must designate a responsible person, called the A/D director, to assemble or disassemble a crane. This individual has legal responsibilities described in 29 *CFR* 1926.1404, *Assembly/Disassembly—General Requirements*. An operator working alone may act as the A/D director, but that individual must be both competent and qualified as defined by OSHA standards.

Workers should rope off an erection area large enough to allow ingress and egress of trucks carrying crane components. Its size should allow layout of the components in proper order for assembly. The perimeter should prevent interference or hazards to other site personnel during crane assembly. In addition, as the crane swings, there needs to be at least two feet of clearance (0.6 m) between the superstructure counterweight and the nearest obstacle.

Before assembly begins, the operator and the person responsible for the lift must consider the following:

- Is there enough room for the crane to maneuver into position on the site?
- Is there a large enough area in which to erect or extend the boom?
- Can the trucks hauling boom sections get into place and be unloaded safely?
- Is there enough timber blocking to support the boom when it is being assembled and disassembled?
- Will crane operations occur within the vicinity of overhead power lines as described in 29 *CFR* 1926.1407–1411?

<table>
<tr><td>**WARNING!**</td><td>Personnel involved in crane A/D work must always follow the manufacturer's guidelines. Never take shortcuts that could lead to injury when climbing or handling boom attachment components. Most are heavy enough to require another crane. Pinching and crushing are common sources of injury.</td></tr>
</table>

Self-Assembling Cranes

Some of today's mobile cranes are designed for self-assembly. Of course, not every task can be automated, such as inserting all the connector pins and their retaining hardware. But cranes such as the Liebherr LR1250 crawler crane can self-assemble the main boom along with a luffing jib after it has been delivered to the jobsite. Videos of the process can be found on sites such as YouTube. Any design innovation that makes the task of assembly easier and safer is welcomed by the industry.

1.3.1 Counterweight Considerations

Everyone involved in crane component assembly and disassembly must be aware of counterweight requirements. When attaching boom extensions, the mounted counterweight places restrictions on the boom-extend mode and superstructure position with some cranes. The crane may overturn if the leverage of the boom-extend mode or position of the superstructure is not within the crane's load chart parameters. Many LMIs can display boom extend-mode information (*Figure 10*).

Workers must always wear the appropriate PPE and fall protection equipment, and observe safe working practices when assembling and disassembling cranes.

Operators must correctly enter the crane's counterweight configuration into the LMI for the system to function properly. All LMI tipping alerts and protective actions rely on accurate configuration inputs. Remember to reprogram the LMI with revised configurations when removing counterweights or changing boom components. LMIs are equipped with an override button to allow assembly operations that would otherwise be prevented by the LMI's programming. The override button must only be used as the manufacturer specifically allows, and never as a matter of convenience to defeat safety programming.

1.4.0 Intermediate Boom Sections

The boom extension of a telescopic crane may include intermediate boom sections (*Figure 11*). This section covers intermediate boom sections and the procedures for installing them on one crane model. Installation on other models is similar, but will differ in the details.

Before installing the boom extensions, determine the boom-extend mode of the crane required for the lift. As with the setup of a swing-away, the operator must extend and set the outriggers, verify the appropriate amount of counterweight, retract the main boom completely, and lower it into the horizontal position.

Only the combinations of boom-extension spans and counterweights indicated in the crane's load chart are permissible. Insufficient counterweight creates the risk of overturning the crane during assembly.

Assembly and disassembly personnel must follow the manufacturer's instructions when installing and removing boom extensions. Use only original equipment manufacturer (OEM) parts, or authorized substitutions with sufficient loadbearing capacity, when assembling boom extensions.

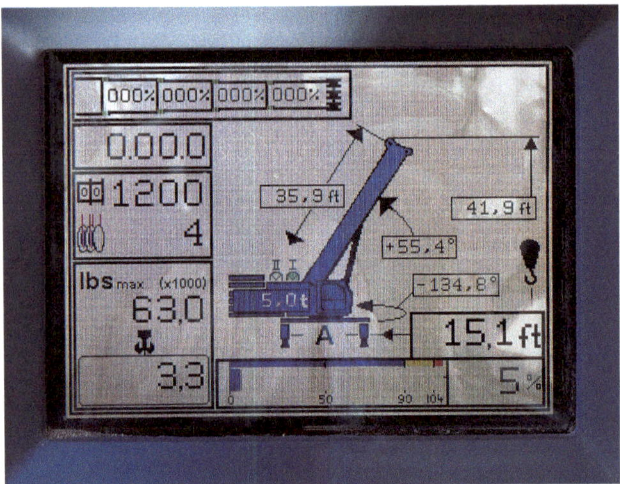

Figure 10 Boom-extend mode on an LMI.

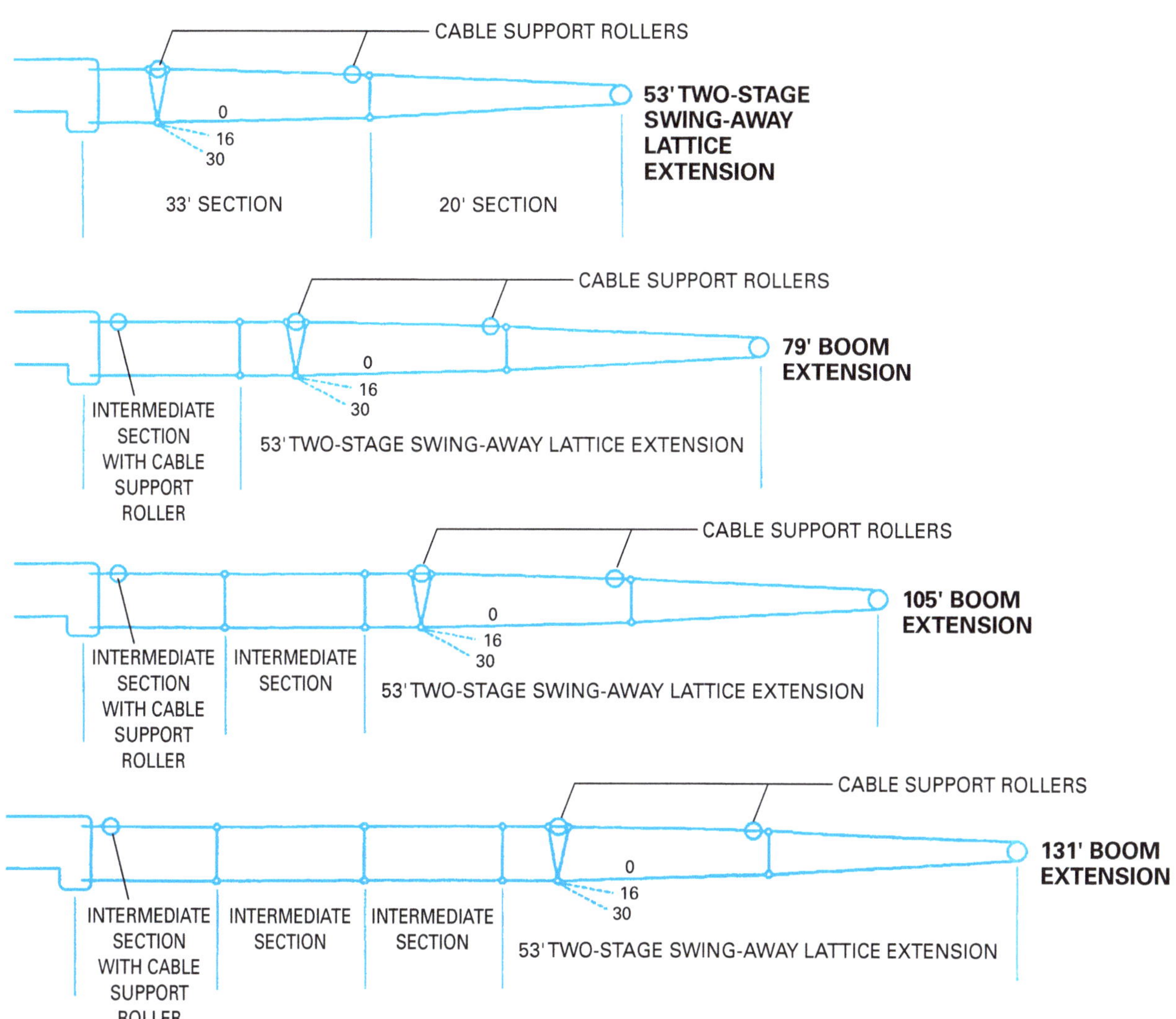

Figure 11 Intermediate boom sections.

1.4.1 Assembling Intermediate Boom Sections

A/D operations usually require an auxiliary crane when attaching intermediate boom sections to the main boom. The following abbreviated procedure is for one specific crane, but the process is similar for other cranes. Always consult the manufacturer's guidelines for the crane in use.

Step 1 Before mounting the intermediate boom sections, remove the swing-away from the main boom. Hook an auxiliary crane to the extension per manufacturer's instructions, hold the swing-away, remove and stow the swing-away connector pins, and place it on the ground.

Step 2 The first intermediate section should have a cable support roller. Attach the lift rigging to the intermediate section. Boom sections include specific attachment points for this purpose.

Step 3 The end of the intermediate section with the cable support roller should be nearest the boom head. Use the auxiliary crane to position the intermediate section so that its anchor points align with the attachment fittings on the boom head.

> **CAUTION**
>
> Always assemble the intermediate section with the cable support roller directly in front of the main boom head. Damage to the hoist rope by the main boom head may occur if the first intermediate section does not include the cable support roller.

> **WARNING!**
>
> Pinch and crush injuries are too common during crane assembly and disassembly. Keep hands clear of attachment points while mating up intermediate boom sections.

Step 1 Remove the pins from the stowage brackets on the intermediate section, and insert them at the top and bottom attach points between the boom head and the intermediate section. Secure the pins using their retaining clips.

Step 2 Use the auxiliary crane to raise and position the swing-away in front of the intermediate section. Align the anchor points on the foot section of the swing-away with the attachment fittings on the intermediate section.

Step 3 Remove the pins from the stowage brackets on the swing-away and insert them at the top and bottom attachment points between the intermediate section and the swing-away. Secure the pins using their retaining clips.

Step 4 If an extension offset is required, use the auxiliary crane to support the end of the assembled extension. Then remove the offset adjustment pins and lower the end of the extension to the desired offset. When the offset pin holes align, then reinsert the pins. Secure the pins with their retaining clips.

> **WARNING!**
>
> Depending on the amount of offset required and the length of the swing-away, the operator may have to raise the boom before attempting to adjust the offset angle. Workers should use fall protection if they need to climb to the offset hinge in such cases.

Step 5 Restore all electrical connections and devices. These may include the anti-two-blocking switch, anemometer, and high-voltage proximity sensors.

Step 6 Place the hoist cable through the anti-two-blocking switch weight, and reeve the hook ball or load block.

1.4.2 Removing Intermediate Boom Sections

When the intermediate boom sections are no longer needed, their removal requires a relatively simple procedure. An auxiliary crane is typically required to disassemble the intermediate sections and make any adjustments to the swing-away offset. Configure the crane as follows for disassembly:

- Retract all telescopic sections completely with the boom raised.
- Turn the superstructure to an appropriate quadrant permitted by the crane's load chart. The amount of space available may be a deciding factor in this choice.
- Lower the main boom until the LMI triggers a shutdown.
- Override the LMI, and lower the main boom into the horizontal position (or until the offset extension can touch the ground, whichever occurs first).

If the boom extension has an offset, proceed as follows:

Step 1 Lower the main boom until the head sheave on the offset boom extension is just resting on a board on the ground. This should exert enough force to leave the offset pins loose.

Step 2 Remove the offset pins from their bearing points.

Step 3 Raise the boom extension to the 0-degree offset position with the auxiliary crane, and pin the extension in this position. Then, proceed with the following steps.

If the boom extension does not have an offset, complete the following steps:

Step 1 Remove the anti-two-blocking switch weight from the hoist cable on the boom extension, and unreeve the ball or load-block hook.

Step 2 Remove the hoist cable from all sheaves on the boom extension.

Step 3 Remove the retaining pin locking the anti-two-block switch from the swing-away.

Step 4 Release the anti-two-block switch lock on the main boom.

Step 5 Disconnect all electrical connections between the distribution boxes at the main boom and the swing-away.

Step 6 Spool the cable onto the appropriate devices.

Step 7 Screw the plugs of the connecting cable onto the dummy sockets of the individual parts of the boom extension.

Step 8 Screw the shorting plugs onto the distribution boxes on the main boom and on the swing-away.

Step 9 Screw the dummy plugs onto the dummy sockets on the intermediate boom sections.

Step 10 Attach the appropriate lifting gear to the swing-away, and hold it with the auxiliary crane.

Step 11 Remove the connector pins between the swing-away and the intermediate boom section, and insert the pins in their stowage brackets on the swing-away. Secure the pins with their retaining clips.

Step 12 Place the swing-away on the ground.

Step 13 Attach the lifting tackle to the intermediate boom section, and hold the section with the auxiliary crane.

Step 14 Remove the pins securing the intermediate boom section, and insert the pins into their stowage brackets on the intermediate boom section. Secure the pins with their retaining clips.

Step 15 Place the intermediate boom section on a transport vehicle.

1.5.0 Luffing Extensions

When the construction site permits, a standard telescopic-crane boom configuration with the swing-away jib may reach all areas of need from one location. Operators can change the load radius within the limitations of the crane's capacity by increasing the boom angle. However, an increase in load radius (and boom tip height) is possible by inserting intermediate sections to extend the jib, lengthening the boom as described earlier. This provides a greater load radius with a smaller boom angle.

In many situations, however, obstructions near the crane site can limit the load radius of a crane using straight-boom configurations. For example, lifting to the roof of a tall structure or hoisting a tall load requires high boom angles. Placing the load may be impossible if the intended location is farther from the crane than the load radius it can achieve at that boom angle.

1.5.1 Manual Luffing Jibs

Cranes can overcome load radius limitations to a certain extent by using a relatively lightweight boom extension at some fixed offset angle. At larger offsets, the crane achieves a greater load radius for a given main boom angle, but at the cost of reduced capacity. Many crane manufacturers provide boom extensions with different lengths to accommodate longer load radiuses.

Setting the offset of a fixed jib—also called a manual luffing jib—requires lowering the boom to the ground. The method for setting the offset varies with the type of extension and the crane manufacturer. Some luffing jibs use offset pins inserted into different pairs of holes to fix the angle of the jib at its base, as described earlier. Other manual jibs consist of the extension hinged-to-base section. Workers set the boom angle by inserting or removing heavy metal links between the two parts opposite the offset hinge (*Figure 12*).

Manually luffing a jib is labor intensive, and requires space to lower the crane boom to make the change. At larger offset angles, the hinge point at the head of the main boom may be many feet off the ground, even with the jib-head sheaves resting on the ground. This increases the fall hazard to workers involved, who will either require a lift or must climb to the jib base to remove and insert the offset pins while making the angle adjustment.

Figure 12 Links used to set a manual luffing jib offset.

1.5.2 Power Luffing Jibs

Several factors have prompted manufacturers to address cost and delay issues related to manual adjustment. Crane users have requested mobile cranes with longer load radiuses at smaller boom angles, but without the need to lay the boom down to adjust the jib offset. Many times, there simply isn't any room to do this. The other factor involves simple physics. As boom angles become smaller and luffing reaches increase, cranes reach both structural-strength and stability limits.

Manufacturers solved the first problem in some luffing jibs by inserting a hydraulic actuator at a hinge point in the jib extension. This feature allows the operator to luff the jib from the cab. The term power luffing jib applies to such boom extensions. Modern crane boom control systems can coordinate the offset with the main boom angle to produce a level luffing action. Since the crane's capacity changes dynamically with such a system, the LMI must be able to integrate the luffing jib's position with the other boom and load parameters.

Solving the second problem required adding one or more jib masts, forestays, and backstays to support longer luffing jibs. Extra-long lattice jibs can more than double the reach of a telescopic crane. The system of stays reduces the tendency of the main boom to buckle by allowing it to operate at high angles. These jibs are luffed by adjusting the lengths of the cable stays and pivoting the jib mast supports (*Figure 13*). In extreme cases, the main boom can be elevated to its highest angle with the jib's backstays connected to a separate set of counterweights behind the crane. In this arrangement, the operator can luff the extension closer to a horizontal position. Operations with long luffing jibs require reference to special load charts and the use of LMIs programmed with these special boom configurations.

Figure 13 Mobile crane equipped with a long power luffing jib.

Cranes and Wind Farms

To get the best value from the investment, wind turbines have grown to incredible heights and rotor sizes. The crane in *Figure 13* is setting a portion of the wind-turbine tower in place. Eventually the nacelle will be installed, shown here in an exploded view. The rotor assembly, with all blades attached to a hub, is often lifted into place fully assembled.

Many crane manufacturers have developed and built cranes specifically designed to accommodate the extremely high lifts of complex structures, while still being able to get in and out of these remote jobsites. Many are large telescopic-boom cranes fitted with complex power luffing jib systems and other accessories to improve ground stability and make the cranes footprint as large as possible. This type of crane takes less time to assemble on site than a comparable lattice-boom model, and fewer trucks are required to transport the boom components.

Figure Credit: Nick Kaloterakis from kollected.com <http://kollected.com>

Additional Resources

ASME Standard B30.5, Mobile and Locomotive Cranes. Current edition. New York, NY: American Society of Mechanical Engineers.

Cranes: Design, Practice, and Maintenance. Second edition. 2002. Ing J. Verschoof. Hoboken, NJ: John Wiley and Sons, Inc.

Crane Handbook. D. E. Dickie. 2013. Waltham, MA: Butterworth-Heinemann.

Mobile Crane Safety Manual. 2014. Milwaukee, WI: Association of Equipment Manufacturers.

29 *CFR* 1926, Subpart CC, *Cranes and Derricks in Construction.* **www.ecfr.gov**

1.0.0 Section Review

1. To luff a boom is to _____.
   a. swing the boom right or left
   b. telescope the boom in or out
   c. raise or lower the boom
   d. shorten or lengthen the load radius by articulation

2. Which of the following statements is *not* true about auxiliary boom heads?
   a. During installation, the workers must plug the boom head's electrical signal cables into the main boom sockets.
   b. Auxiliary boom heads can handle up to ten parts of line.
   c. Workers must install the rope guide bars after reeving the hoist cable through the sheave(s).
   d. Leading the hoist cable through the anti-two-block weight is one of the last steps when installing an auxiliary boom head.

3. The legal responsibilities for an A/D director are found in _____.
   a. *ASME B50.5—Mobile and Locomotive Cranes*
   b. *ASME P30.1—Planning for Load Handling Activities*
   c. 29 *CFR* 1926.32
   d. 29 *CFR* 1926.1404

4. When installing the first intermediate boom extension section, its cable support roller must be _____.
   a. on the end nearest the main boom head
   b. on the opposite end from the main boom head
   c. on the underside of the section
   d. a double-sheave roller

5. Which option provides the greatest allowed load radius for a telescoping-boom crane equipped with a manual luffing jib?
   a. Setting the jib with a 0-degree offset
   b. Raising the boom
   c. Retracting the boom
   d. Setting the jib with the largest offset angle possible

Summary

Lifting operations depend on the proper setup and assembly of the crane boom. On telescopic cranes, this can involve the attachment of jibs or intermediate boom extensions to the main boom. While some tasks, such as deploying a fixed foldaway jib, are considered crane setup, other tasks fall into the category of assembly and disassembly. These tasks require an A/D director to ensure they are performed safely.

Workers must be aware of clearance requirements on the jobsite and follow the safety requirements associated with setup and assembly/disassembly work. The procedures presented in this module should be considered general guidelines; always follow the manufacturer's instructions when performing setup and assembly for a specific crane.

1. A significant factor contributing to accidents during mobile crane assembly/disassembly is _____.

    a. incomplete manufacturer's instructions
    b. using worn parts
    c. human error
    d. weather

2. During setup of a swing-away extension, the purpose of the tag line is to _____.

    a. secure the swing-away until ready to position it
    b. mark on the ground the extent of the extension's swing to keep spectators clear
    c. string together all the connector pins to prevent losing them
    d. control the motion of the swing-away during setup

3. When stowing a swing-away extension, the step listed below that should be completed first is to _____.

    a. disconnect and secure all extension electrical connections
    b. remove the connector pins from the extension base
    c. stow the stop rail in the transport position
    d. wind in the hoist rope

4. When a crane does not have an attached auxiliary boom head that can be deployed at the end of the main boom, _____.

    a. an auxiliary boom head cannot be used at all
    b. an auxiliary crane must lift the component into position
    c. the boom will have to be significantly field-modified
    d. the additional weight of adding one will cause instability

5. The amount of counterweight required for a given boom-extend mode is determined by the _____.

    a. A/D director
    b. load moment indicator (LMI)
    c. crane operator
    d. crane's load chart

6. Usually, the swing-away for a telescoping-boom crane is set up on the last intermediate extension section by _____.

    a. rotating it into position from its stowed location
    b. lowering the boom tip to the ground and attaching the swing-away there
    c. using an auxiliary crane to lift the swing-away into position
    d. having a crew of workers lift the swing-away into position

7. When disassembling a boom extension that includes intermediate sections and a swing-away with an offset, workers must first _____.

    a. remove the connector pins from the swing-away and place it on the ground
    b. use an auxiliary crane to set the swing-away offset to 0 degrees
    c. place the intermediate sections on transport vehicles
    d. remove the connector pins from the first intermediate section

8. Before removing any type of boom extension from a telescoping-boom, it is not necessary to _____.

    a. reprogram the LMI
    b. remove the hoist cable from all sheaves in the extension
    c. disconnect all electrical connections between the boom and swing-away
    d. remove the anti-two-blocking switch and weight from the swing-away

9. Which limitation in crane operations led to the development of longer power luffing jibs?

    a. Low ground-bearing pressures
    b. Too-small outrigger footprints
    c. Crane instability at long load radiuses and small boom angles
    d. Hoist ropes with too-low single-line pull capacities

10. Which feature do all power luffing jibs have?

    a. Extension offsets are set with pins.
    b. Offset angles are controlled remotely by the crane operator.
    c. Luffing angles are set by moving the main boom.
    d. They are all the telescoping type.

Trade Terms Introduced in This Module

A/D director: The individual specified by OSHA standards to be responsible for the assembly and disassembly of cranes.

Boom-extend mode: Refers to the order, and percentage, of extension telescopic-boom sections. The standard mode is to extend all segments the same percentage. Other modes may selectively extend certain sections to provide different capacities for a given load radius.

Connector pin: Any solid, cylindrical, metal part used to fasten, support, or attach the various components and members of a crane's boom and other hoisting gear.

Fly: With reference to mobile cranes, a boom extension carried on the side of the main boom when not in use, that can swing into position to extend the main boom; also called a swing-away.

Level luffing: The process of luffing the boom of a crane while adjusting the hoist cable and boom extension offset so that the load travels in a horizontal direction.

Luffing: Raising or lowering a crane's boom. A luffing jib can luff independently of the main boom.

Manual luffing jib: A semi-fixed jib extension whose offset must be manually set between lifts.

Power luffing jib: A jib extension that has the means to change its offset during a lift by remote control of the operator.

Stop rail: A heavy metal guiderail used by some crane manufacturers to help align a swing-away to its transport location as it is being stowed.

Additional Resources

This module is intended as a thorough resource for task training. The following reference materials are recommended for further study.

ASME Standard B30.5, Mobile and Locomotive Cranes. Current edition. New York, NY: American Society of Mechanical Engineers.

Cranes: Design, Practice, and Maintenance. Second edition. 2002. Ing J. Verschoof. Hoboken, NJ: John Wiley and Sons, Inc.

Crane Handbook. D. E. Dickie. 2013. Waltham, MA: Butterworth-Heinemann.

Mobile Crane Safety Manual. 2014. Milwaukee, WI: Association of Equipment Manufacturers.

29 CFR 1926, Subpart CC, *Cranes and Derricks in Construction*. **www.ecfr.gov**

Figure Credits

© Eugenesergeev/Dreamstime.com, Module Opener

Link-Belt Construction Equipment Company, Figures 1, 2

The Manitowoc Company, Inc., Figures 3–8, 11

© iStock.com/gmalandra, Figure 9

Liebherr USA, Co., Figure 12

© iStock.com/axel-ellerhorst, Figure 13

Nick Kaloterakis from kollected.com <**http://kollected.com**>, SA01

Answer	Section Reference	Objective
Section One		
1. c	1.1.0	1a
2. b	1.2.0	1b
3. d	1.3.0	1c
4. a	1.4.1	1d
5. d	1.5.1	1e

Lattice Boom Assembly and Disassembly

OVERVIEW

The long length of some lattice booms makes it impossible for them to be transported when assembled. Therefore, many lattice-boom cranes require assembly at the work site. Boom assembly and disassembly can be extremely hazardous. Due to the weight and size of boom components, workers must take care when handling them and remain safety-conscious at all times. This module presents the general procedures for lattice-boom assembly and disassembly.

Module 21306

Trainees with successful module completions may be eligible for credentialing through the NCCER Registry. To learn more, go to **www.nccer.org** or contact us at 1.888.622.3720. Our website has information on the latest product releases and training, as well as online versions of our *Cornerstone* magazine and Pearson's product catalog.

Your feedback is welcome. You may email your comments to **curriculum@nccer.org**, send general comments and inquiries to **info@nccer.org**, or fill in the User Update form at the back of this module.

This information is general in nature and intended for training purposes only. Actual performance of activities described in this manual requires compliance with all applicable operating, service, maintenance, and safety procedures under the direction of qualified personnel. References in this manual to patented or proprietary devices do not constitute a recommendation of their use.

Lattice Boom Assembly and Disassembly

Objectives

When you have completed this module, you will be able to do the following:

1. Identify boom components and primary considerations in the assembly and disassembly of lattice booms.
   a. Identify primary lattice-boom components.
   b. Identify pre- and post-assembly considerations.
2. Explain how to assemble and disassemble lattice booms and jibs.
   a. Explain the general method for assembling lattice booms.
   b. Explain the precautions and considerations for assembling and raising long lattice booms.
   c. Explain how to assemble jibs.
   d. Explain how to disassemble lattice booms.

Performance Tasks

Under the supervision of your instructor, you should be able to do the following:

1. Identify lattice-boom components.
2. Determine the amount of counterweight required for lattice boom erection.
3. Determine the proper boom-section count and their layout for a specific lattice boom.

Trade Terms

Back hitch
Belly lines
Boom stops
Cantilever length
Diagonal struts
Gantry

Hoist kick-out
Lattice lacing
Live boom mast
Main chords
Pendant
Suspension rope

Industry Recognized Credentials

If you are training through an NCCER-accredited sponsor, you may be eligible for credentials from NCCER's Registry. The ID number for this module is 21306. Note that this module may have been used in other NCCER curricula and may apply to other level completions. Contact NCCER's Registry at 888.622.3720 or go to **www.nccer.org** for more information.

Contents

Figures

1.0.0 LATTICE BOOMS AND ASSEMBLY CONSIDERATIONS

Objective

Identify boom components and primary considerations in the assembly and disassembly of lattice booms.

a. Identify primary lattice-boom components.

b. Identify pre- and post-assembly considerations.

Performance Tasks

1. Identify lattice-boom components.

2. Determine the amount of counterweight required for lattice boom erection.

3. Determine the proper boom-section count and their layout for a specific lattice boom.

Trade Terms

Diagonal struts: In a lattice boom with a rectangular cross section, pieces of structural pipe or angle iron welded to diagonally opposite chords of the boom to stiffen the assembly.

Gantry: A heavy frame with roller sheaves mounted on the upperworks of a crane, used for providing a mechanical advantage to the boom hoist ropes; also refers to a jib strut or mast.

Lattice lacing: In a lattice boom, the zig-zag pattern of structural metal supporting the main chords along the sides, top, and bottom of the boom sections.

Main chords: The heavy structural pipe or angle iron members running the length of a boom section.

Pendant: A length of heavy wire rope or flexibly connected sections of flat metal bars, equipped with connectors at both ends, used to hoist and/ or support sections of a crane's boom.

Suspension rope: The rope reeved through a crane's gantry and mast sheaves to the sheaves suspending the boom pendants.

Depending on a crane's size and weight, assembly and disassembly (A/D) workers must disassemble cranes for transportation and reassemble them at new work sites. For most lattice-boom cranes, A/D workers remove the boom and counterweights before transport. For larger cranes, they also remove the crawler treads, because the carbody with crawlers is too wide for standard road lanes. Once the crane arrives at the site, A/D personnel and operators must unload and assemble the crane and its boom. *Figure 1* illustrates the general assembly process for one model of crane, though the steps are similar for most lattice-boom cranes.

Assembling cranes can be hazardous. The results of a two-decade study ending in 2006 indicate that over half of all fatalities from impact by falling crane boom components were the result of unskilled workers setting up cranes. OSHA has addressed the particularly hazardous nature of crane assembly and disassembly in 29 *CFR* 1926.1403 through 1926.1406. Some of these requirements include the following:

- An individual called the A/D director, who meets OSHA's criteria for being both a competent and qualified person, or a competent person assisted by qualified persons, must supervise all A/D operations.
- A/D personnel must understand their tasks and the hazards associated with crane assembly.
- The operator and A/D personnel must communicate and coordinate actions to avoid movement of equipment when workers are out of the operator's sight.
- The A/D director must address crane ground support, load characteristics, and other planning considerations associated with handling crane components.
- Before use, operators must test any hoist or swing brake they intend to rely upon to prevent crane movement. Otherwise, the operator must engage a pawl or install a locking device on the drum or gear, or workers must use blocking or an assist crane to prevent movement.
- A/D personnel must know the weights of all components they will handle to ensure the stability of the crane they are assembling, and to remain within the capacities of assist cranes.

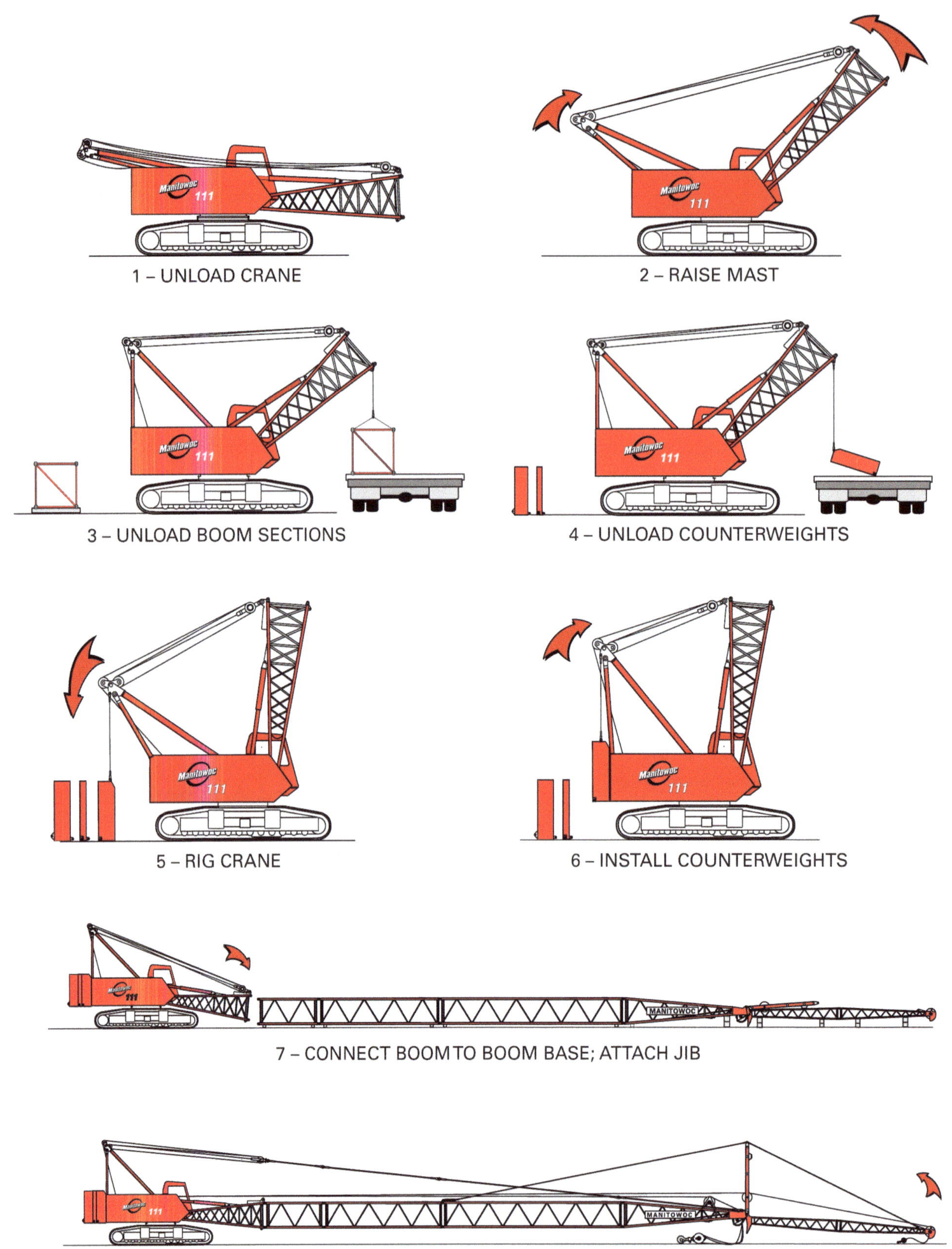

Figure 1 Crane assembly process.

- A qualified person must conduct a post-assembly inspection before lifting operations may begin, in accordance with 29 *CFR* 1926.1412, *Inspections*.
- Qualified riggers must perform all A/D rigging work.
- Workers shall not partially or completely remove pendant pins when the pendants are in tension (*Figure 2*).
- Workers shall not remove boom connector pins between the pendant attachment points and the crane body when the pendants are in tension (*Figure 3*).
- Workers shall not remove any boom connector pins when only blocking under the boom tip supports the boom, or the upper pins in the cantilevered section of the boom (*Figure 4*).

Crane operators must help to ensure crane assembly is done correctly according to the manufacturer's instructions. The project engineer, site superintendent, contractor, and the user of the crane are also responsible for safety.

Specific guidelines and instructions for assembling the crane and its boom are provided in the manufacturer's documentation for each crane. Some manufacturers may provide a separate A/D manual. *Figure 5* shows the table of contents for a typical set of assembly and disassembly procedures.

1.1.0 Boom Sections

Lattice booms, especially large ones with various jibs added, can be comprised of many independent sections. After lift planners have determined the boom model and configuration needed to accomplish the lift, the necessary pieces must be collected from inventory and prepared for shipment. The pieces required can be determined from a crane's boom combination chart in the documentation. An example of such a chart for the Manitowoc 18000 can be reviewed in the *Appendix*. The components should be inspected before they are loaded for shipment to avoid hauling damaged components to the site.

After the boom sections are unloaded at the site, they should be inventoried and inspected again. The crane's rigging drawings shows the arrangement of boom sections and their names or identification numbers. The A/D workers should inspect the boom sections for damage or wear. Qualified crane repair workers must repair or replace damaged components according to the manufacturer's specifications.

Damaged or worn components can cause serious injuries. The boom sections should be checked for bent main chords, lattice lacing or diagonal struts (*Figure 6*), as well as any failing welds. A/D workers can't fit the boom sections together if there are bent or damaged fittings. Components that are damaged or deformed in some way should be rejected. When alignment problems occur consistently, it often indicates that one or more components are damaged. Check that the gantry or mast suspension rope, pendant sections, and boom pendant pins are all in good condition. Connector pins must have spring-steel retaining clips or cotter pins, as shown in *Figure 7*. Cotter pins should never be reused, but retaining clips and pins provided by the manufacturer are intended for reuse.

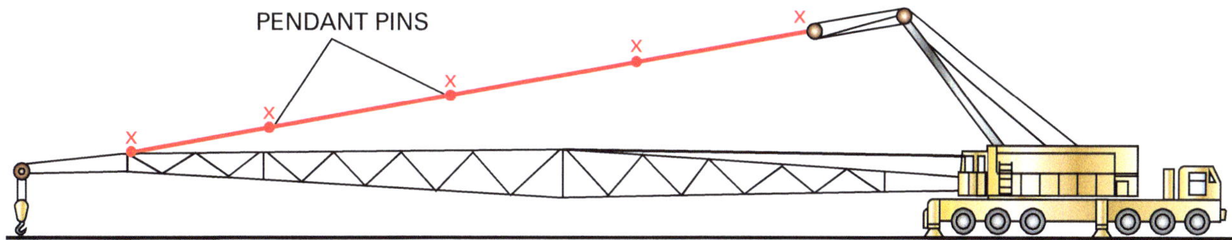

Figure 2 Never remove pendant pins when the pendant is under tension.

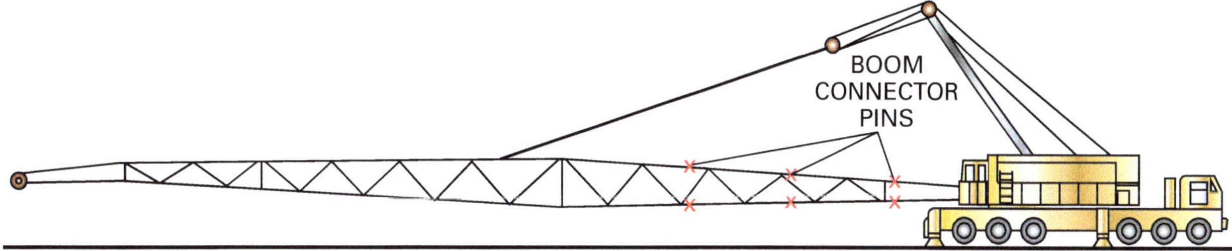

Figure 3 Never remove boom connector pins between the pendant under tension and the boom base.

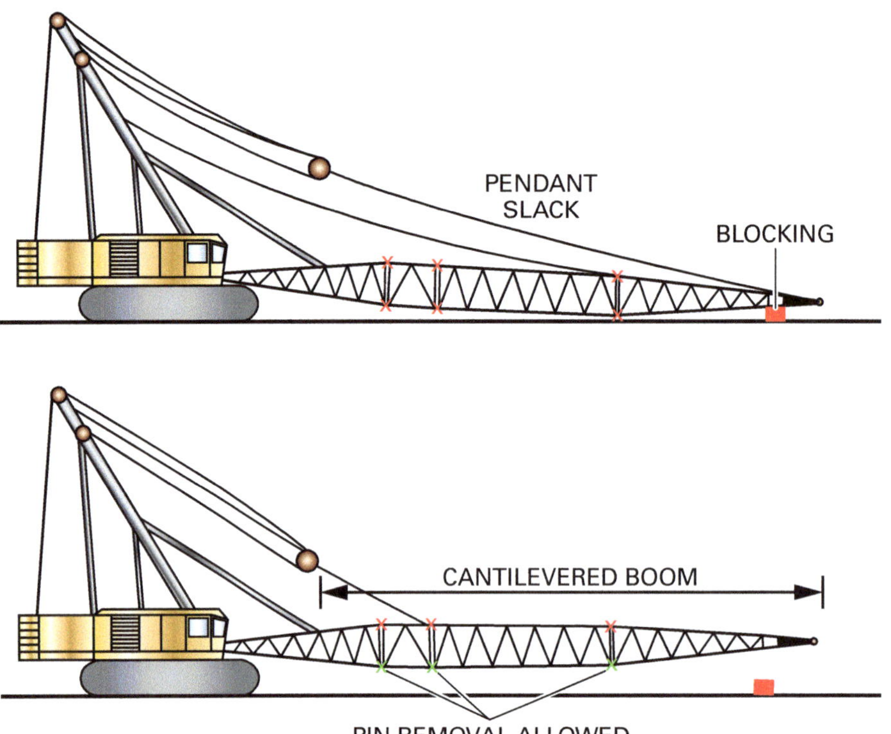

Figure 4 Never remove boom connector pins when the boom is supported only at its end or the connectors are under tension.

Case History

Lattice Boom Collapse During Disassembly

In 2008, an adult male died when crushed while assisting in the disassembly of a lattice boom. The boom consisted of a 15-foot base section, two intermediate boom extensions (20 feet and 10 feet), and a 10-foot tip that included the suspended hook block. Four connector pins and eight bolts secured adjacent intermediate sections.

The worker was a friend of the owner and had no formal training with assembling cranes. The owner had purchased the used crane, and he also had no formal crane A/D training. After positioning the boom near the horizontal with an auxiliary crane, the two men proceeded to remove the bolts and pins for the outermost sections. The worker reported that two of the pins seemed stuck. When he struck one of the pins with a sledge hammer, the boom collapsed, crushing him.

Lessons learned from the OSHA investigation of this accident include the following:

- Workers must be trained for crane disassembly operations.
- Crane disassembly procedures must be followed as written.
- Always properly support all boom sections before beginning disassembly.
- Never take a position underneath or inside of boom components during assembly or disassembly.
- Use a long bar to remove pins from outside the boom; never climb inside to remove retaining clips or to remove connector pins.
- Workers should be trained to insert boom assembly pins from the inside of the boom so the retaining clips are on the outside for easy access.

Table of Contents

Component Identification

NOTE: Except for procedures specific to #82 boom, mast for #22E boom is shown throughout this folio. Mast for #22EL and #82 booms is identical except that pendants (#22EL boom) or straps (#82 and #82LR boom) are suspended from mast.

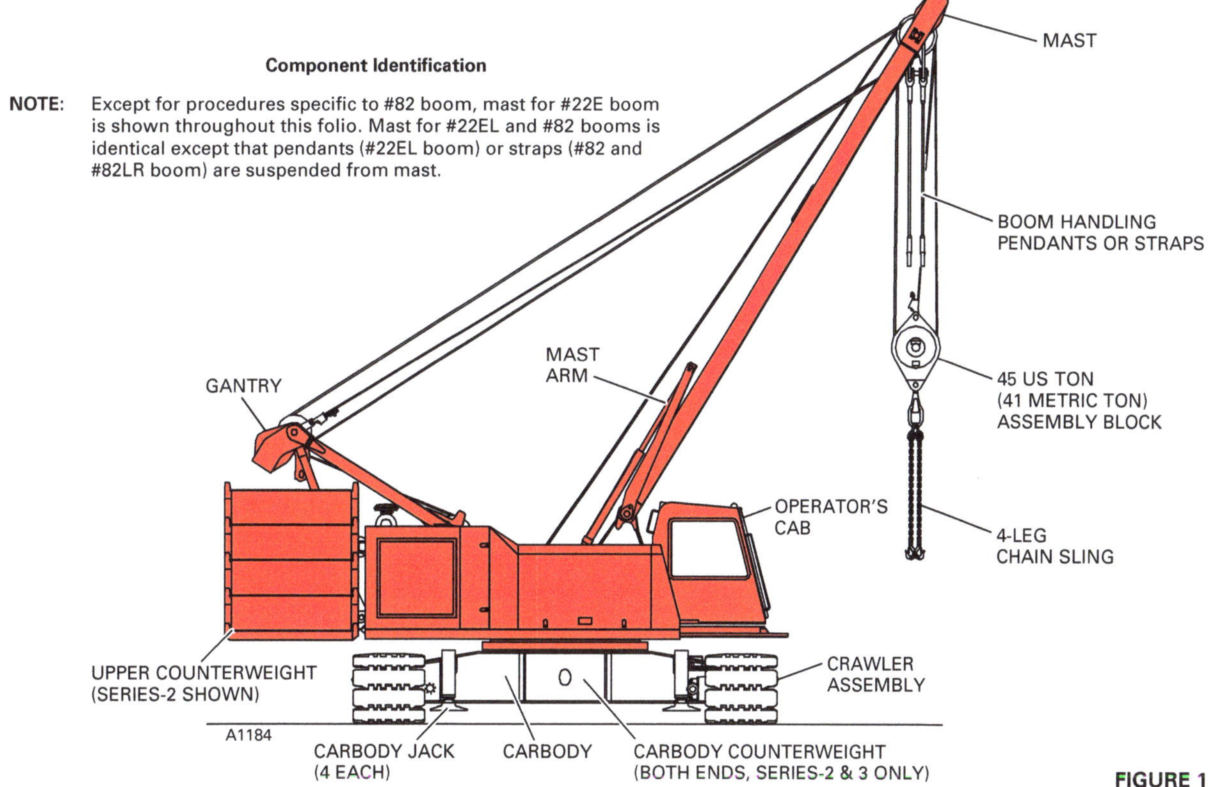

© 2002 MANITOWOC CRANES, INC.

Figure 5 Crane assembly procedure table of contents.

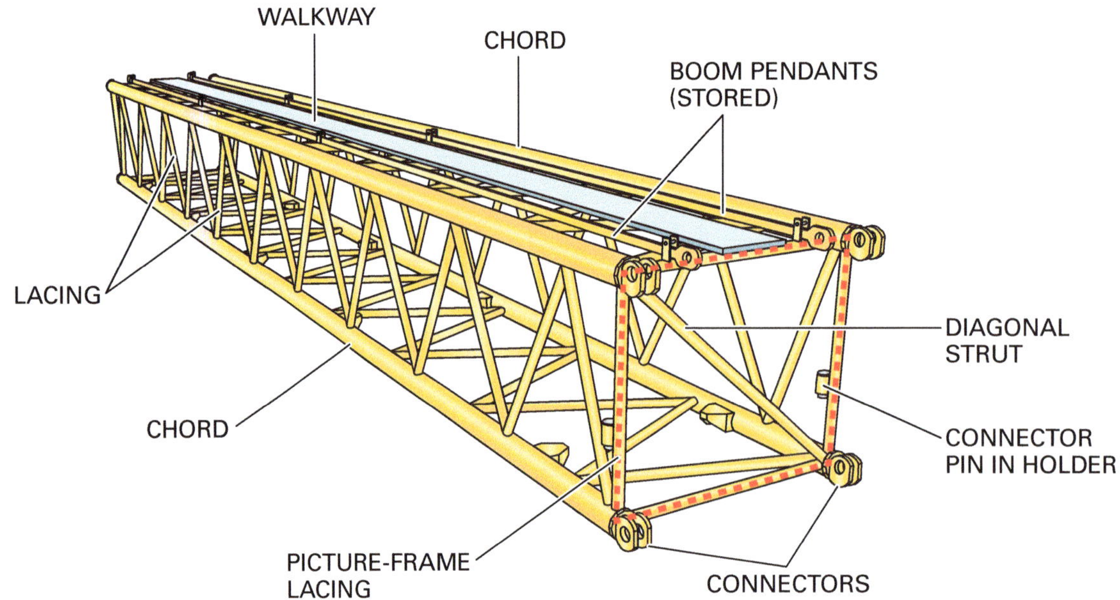

Figure 6 Construction of a lattice-boom section.

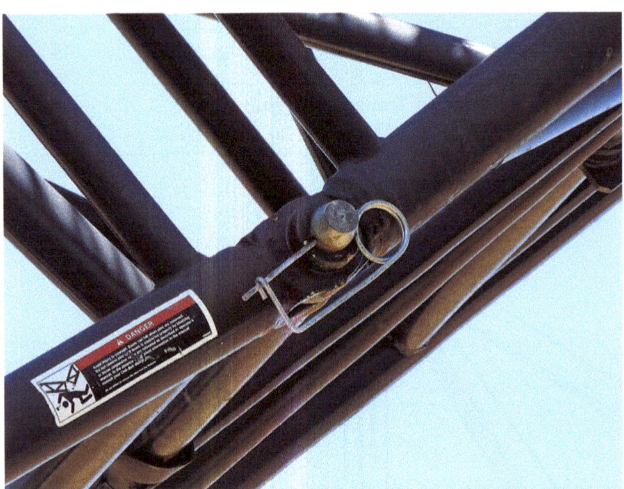

Figure 7 Connector pin secured with a reusable retaining clip.

Before assembly begins, lift planners should have determined the weight of the load(s) and rigging for the task, the height of the lift, and the maximum load radius required. While different loads may be handled, planners must give significant consideration to the lift that will place the most stress on the crane. They will normally use the crane's load chart to select the shortest boom height and radius dimensions that will accomplish the planned work.

Before beginning the assembly, A/D workers and operators should determine the size and quantity of boom connector pins needed to secure the sections. They will need to inventory the required components and replace any damaged or missing items with components meeting the original manufacturer's specifications.

Boom sections must be handled with care to prevent damage to chords, lacing, and connectors. It is important to pay close attention to the type of slings used for the sections. Synthetic slings are generally best, although workers may use other rigging. When using wire or chain slings, A/D workers need to keep the slings clear of the boom lacing to avoid placing stress on these components. Many boom sections include lifting lugs, as shown in *Figure 8*. If this is the case, workers should attach bridles or slings to the lugs.

1.2.0 Pre- and Post-Assembly Considerations

Before assembling lattice booms, a good assembly location must be determined. The area should have a firm, level, uniform supporting surface. It must be large enough to accommodate the crane, the length of the assembled boom, and the supporting truck traffic. A barricade should separate the assembly area from public areas. The barricade should not interfere with, or pose a hazard to, the task and site personnel.

The assembly area should be large enough to allow staging and handling the largest components. As the crane swings, there must be at least 2 feet (0.6 m) of clearance between the counterweight and the nearest obstacle. *Figure 9* shows an assembly area for a large crane, set apart from other construction activities.

The A/D director, operator, lift planners, and persons responsible for the intended lifts must take the following logistics into consideration:

- Is there enough room for the crane to maneuver into position?
- Is there sufficient room to assemble the boom?
- Can the trucks hauling boom sections get into place and be safely unloaded?
- Is there enough timber blocking to support the boom during assembly and disassembly?

Figure 8 A boom lifting lug.

Figure 9 Assembly area for a large crane.

- Is there an effective system for the crane operator to communicate with the A/D workers?

1.2.1 Counterweight Considerations

Crane operators and any personnel involved in the assembly of lattice-boom cranes must be aware of counterweight requirements. Before assembly begins, verify that the correct counterweight sections required to lift the boom after assembly are available and installed. Lift planners should also verify that the counterweight is sufficient for the lifting operation(s).

Never add or remove a counterweight from a carrier-mounted crane without first positioning the machine on firm, level ground, and extending all outriggers, as shown in *Figure 10*. To lift longer booms off the ground, it may be necessary to install additional counterweight sections. Workers may add these to the standard counterweight on the upperworks, on the sides of the crawler assemblies, or in the bumper areas of the carrier. For some cranes, workers must remove this extra weight after the boom is raised; too much weight behind the point of rotation can reduce backward stability. Always refer to the crane's load chart for allowable counterweight configurations.

<table>
<tr><td>WARNING!</td><td>To avoid tipping the crane, install counterweight sections in the exact sequence shown in the manufacturer's documentation. Ensure that all counterweight components are properly secured; if they move or fall unexpectedly, they can cause serious injury or death.</td></tr>
</table>

Figure 10 Installing counterweights.

Additional Resources

ASME P30.1, Planning for Load Handling Activities. Current edition. New York, NY: American Society of Mechanical Engineers.

Mobile Crane Safety Manual. 2014. Milwaukee, WI: Association of Equipment Manufacturers.

29 *CFR* 1926, Subpart CC, *Cranes and Derricks in Construction*. **www.ecfr.gov**

1.0.0 Section Review

1. Connector pins in pendants should never be removed _____.
   a. unless the pendant is under light tension
   b. without first removing the pendant from the crane
   c. until the proper lubricant has been applied
   d. when the pendant is under any tension

2. Which of the following features is necessary for a crane assembly area?
   a. Soft, sandy soil for cushioning the boom sections
   b. Easy entry and egress for transport trucks
   c. Gently sloping area for assembling the boom
   d. Minimum of 220 VAC electrical service

2.0.0 LATTICE BOOM AND JIB ASSEMBLY

Objective

Explain how to assemble and disassemble lattice booms and jibs.

a. Explain the general method for assembling lattice booms.
b. Explain the precautions and considerations for assembling and raising long lattice booms.
c. Explain how to assemble jibs.
d. Explain how to disassemble lattice booms.

Trade Terms

Back hitch: The mechanism that permits adjusting the position of the boom gantry to maximize its leverage on the boom hoist cables.

Belly lines: Additional pendants attached to the boom hoisting bail and secured at a location near the midpoint of a long boom to minimize sagging.

Boom stops: Devices that mechanically prevent the boom from exceeding a maximum boom angle, reducing the potential for backward tipping or structural damage.

Cantilever length: The length of a horizontal boom that projects beyond the support of the boom pendants, leaving that portion of the boom supported at only one end, like a cantilever.

Hoist kick-out: A term for any safety device that automatically stops a hoist drum when a limit is reached.

Live boom mast: A heavy framework or lattice boom with sheaves, hinged at the lower end of its base on the crane's upperworks, whose purpose is to control the angle of the main boom through attached boom pendants. Boom hoist cables reeved through the boom gantry sheaves or hydraulic actuators control its position.

NOTE

The information presented in this module is general in nature because there are numerous crane models and manufacturers. Always refer to the manufacturer's documentation for the specific crane in use before beginning any assembly work.

Manufacturers provide a boom rigging drawing with each boom ordered. The drawings can include details about boom, boom gantry, back hitch, boom-point, and pendant assembly. The operator should study and understand the content of these drawings before beginning the assembly.

WARNING!

Boom section connection points can be potentially hazardous pinch points. Avoid placing the fingers and limbs into these areas when an unstable condition exists.

2.1.0 Assembling Lattice Booms

This section presents a general procedure for assembling lattice booms. Always consult the manufacturer's documentation before any assembly or disassembly takes place.

2.1.1 Initial Crane Setup

Before beginning assembly, the assembly team must review the planned crane configuration. It is essential to know the model and length of the boom to be assembled. With this information, the operator and A/D director can verify, from the crane's load chart, the amount of counterweight required. Ideally, there needs to be enough counterweight to safely lift the boom off the ground unassisted. However, this is not always the case.

Depending on the type of carrier, assembly may include installation of the crawler assemblies. This step generally involves erecting the gantry system and often a live boom mast as specified by the manufacturer. Initial setup of a basic crawler crane will generally proceed in the following order:

1. Deploy outriggers for supporting the carbody/upperworks
2. Erect the gantry and boom mast rigging
3. Install crawler assemblies
4. Set down on crawlers
5. Install the required counterweight sections

An attempt should be made to position the crane so that the upperworks align with the crane's most stable quadrant pointing in the direction of boom assembly. The boom sections can then be laid out. This may not be possible, however, due to space restrictions. Every jobsite is unique. It may be easier to lay out the boom sections first, then travel the crane into position.

Some crane models have an extendable counterweight that can increase the crane's leverage. Manufacturers may require extending a movable counterweight for boom assembly and disassembly.

2.1.2 Boom Assembly

After positioning the crane over the boom base section according to the documentation, set the swing lock to secure the upperworks. Extend the gantry or mast (*Figure 11*), if applicable, to install the boom base. A mast with special pendants, as shown in the figure, permits the crane to self-assemble the boom base section without the need for an assist crane.

> **NOTE**
>
> Crane load charts or manuals may specifically address the use of the gantry or live boom mast for assembly operations. If required, workers must set up these components first before attempting to assemble and raise a boom. Crane manufacturers will specify the maximum length of boom a crane can raise using its standard equipment. Longer booms require the use of special equipment.

Using pendants or slings as required, raise the hinge connectors of the boom base into position and insert the boom hinge pins. On many models, the pins are hydraulically actuated due to their size. After connecting the boom base hinge, attach the initial set of boom pendants (*Figure 12*). Then install the boom stops, if applicable.

> **WARNING!**
>
> A/D workers must install boom sections and components using the sequence and methods specified by the manufacturer. A boom can collapse if not properly assembled. Do not mix boom sections from different crane models or manufacturers.

With the boom base installed, workers can proceed with assembling the boom sections in the order specified by the manufacturer's manual. The boom number may be stamped on the side of the boom joint connectors or on the diagonal struts of adjacent boom sections. The boom may also have an identification plate. Boom identification is done to prevent mixing boom sections of one type with those of another. Boom identification numbers can be in a variety of formats.

The A/D director must ensure that workers lay down the boom sections in the order of assembly. Generally, manufacturers recommend placing the shorter boom sections closest to the boom base. For shorter booms, workers may pre-assemble the boom and boom head as a unit before attaching it to the boom base.

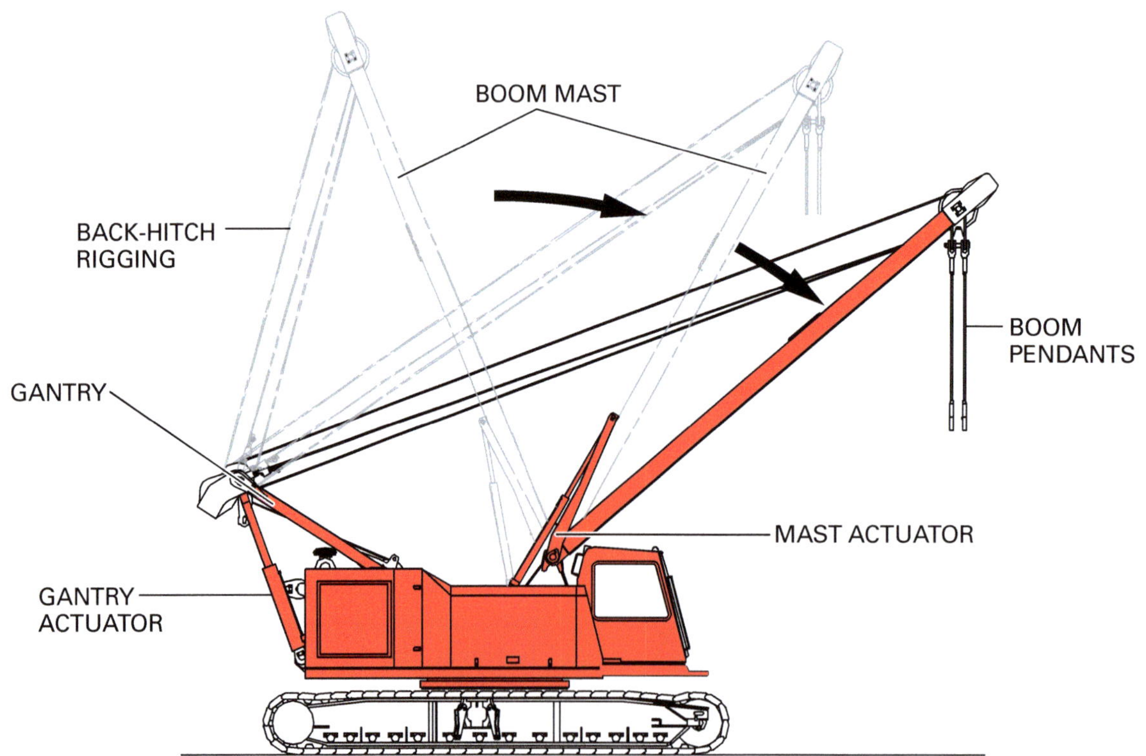

Figure 11 Extending the live boom mast.

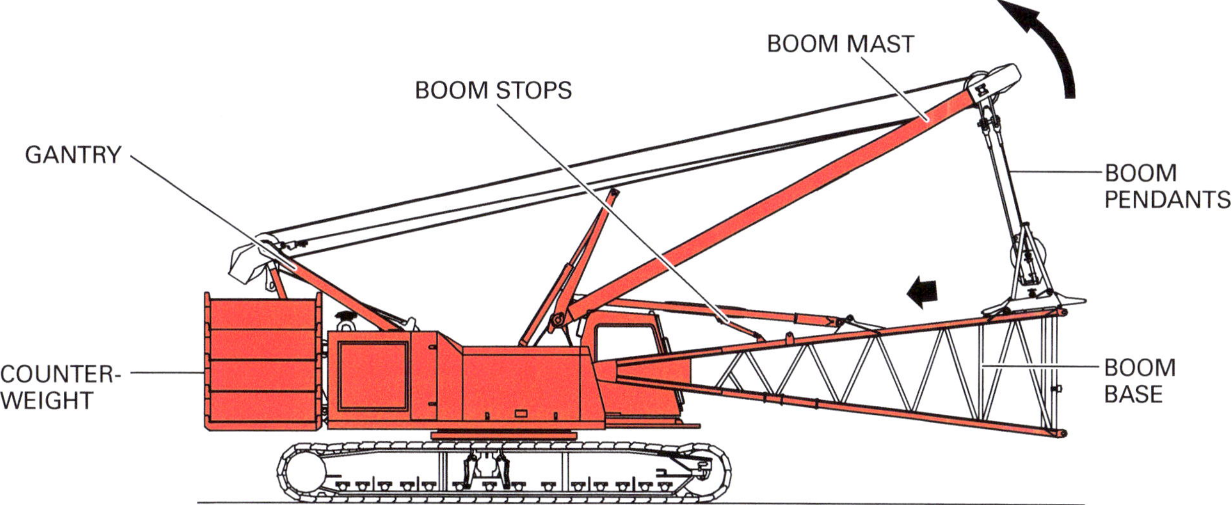

Figure 12 Installing the boom base.

Travel the crane to position the end of the boom base above the first intermediate boom section, also called an *insert*. Lower the boom base to the level of the first boom section. Align the upper connector holes between the two sections being pinned together. Insert the two upper connector pins and their pin retainers or cotter keys (*Figure 13*).

> **WARNING!**
>
> Do not stand inside or under the boom while installing or removing connector pins. Always stand outside the boom unless the employer determines that there is no other way to accomplish the task. In such a situation, the A/D director must implement positive measures to prevent equipment movement when performing these tasks. *Appendix B of 29 CFR 1926, Subpart CC, Assembly/Disassembly: Sample Procedures for Minimizing the Risk of Unintended Dangerous Boom Movement*, covers these situations.

Boom up slightly until the lower pin connection points align. Engage the boom or gantry/mast hoist pawl. Install the pins and retainers (*Figure 14*).

Typically, assembly procedures require attaching pendants to the outermost section of the assembled boom that will connect to the next section lying on the ground. This approach provides the maximum leverage for lifting the boom without assistance. Pendants come in matched sets of either two or four. Builders usually stamp identification numbers into both ends of each pendant to aid in keeping them in sets. Always install pendants in the sequence noted on the rigging drawing.

> **WARNING!**
>
> A set of pendants must be exactly the same length. If they are not, the boom will twist, creating the potential for catastrophic boom failure.

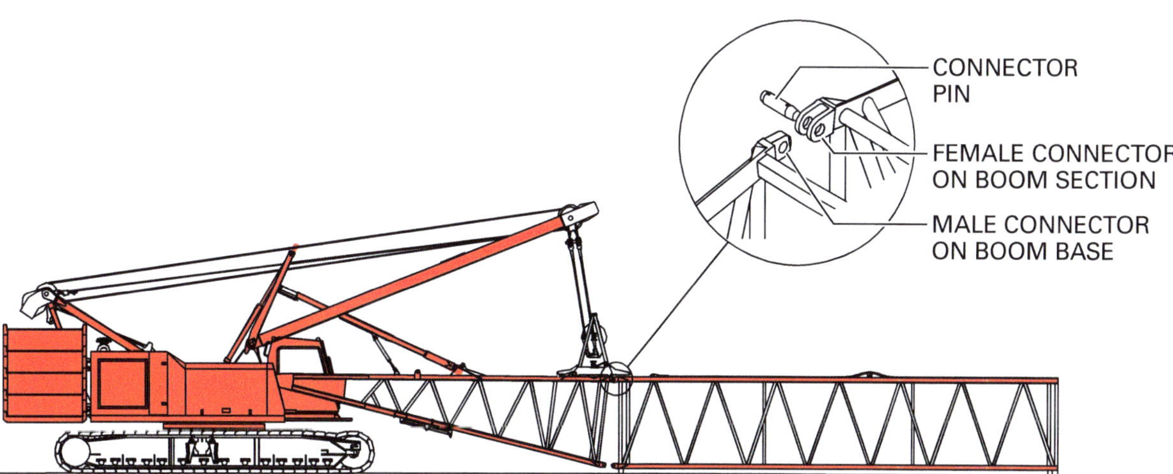

Figure 13 Install the upper connector pins.

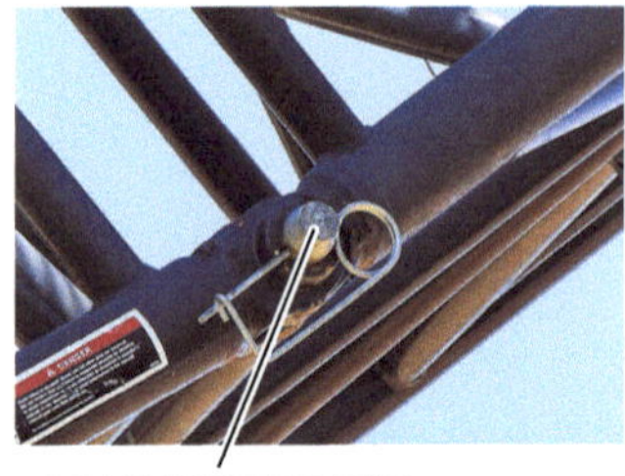

Figure 14 Installing lower pins.

Manufacturers of lattice booms often supply pendants attached to each boom insert for their cranes. With the boom end resting on blocks, workers extend the pendants to the last attached section by detaching the previous set of pendants. The workers then attach these to the free ends of the next set of pendants using pendant assembly pins and retainer clips.

> **CAUTION**
>
> When assembling the boom, make sure that slack suspension ropes or pendants do not snag the connector pins and/or their retainers.

> **WARNING!**
>
> Check the crane's load chart before raising the boom. It may be necessary to block the crawlers, if applicable, to maintain stability.

Assembly of the boom continues in the same manner—join the upper connectors of the next insert, slightly raise the boom to join the lower connectors, lower the boom onto blocks, and move the pendants to the last assembled insert. Repeat until workers have attached all the inserts.

Installing the boom head section is similar to attaching an insert. If the crane's instructions permit, workers may attach the head section to the boom resting on blocks before attaching the boom to the base as previously described. If separately assembling the boom head, the operator first lifts the assembled boom, sets the appropriate hoist drum pawls, then travels the crane to the head section resting on blocks. Alternatively, an assist crane can place the head section at the end of the assembled boom on blocking. The operator adjusts the boom so that the upper connectors align. After inserting and securing the upper connector pins, the operator booms up slightly until the lower connectors align. Workers then insert the lower pins.

After attaching the head section, extend the pendants to the head, if the manufacturer requires. Then boom up so that workers can place blocking under the boom head (*Figure 15*) for completing the remainder of the assembly. Many boom heads include a special support bracket or foot for this purpose.

> **CAUTION**
>
> If the pendants are too long for a crane equipped with a shorter boom and a live boom mast, the mast may tip backwards when approaching maximum boom angle. The weight of the tipping mast can pull an unloaded boom into the boom stops. The operator will have to pull the boom away from the stops by loading the hook.

The remainder of the boom assembly process involves preparing the boom for lifting operations. Workers should make up any electrical connections needed between the crane base and the boom head (*e.g.*, the anti-two-blocking circuit). The cabling required is typically stored on a reel to accommodate varying boom lengths (*Figure 16*). Then the main hoist line is rigged as well as the whip line, if needed. Depending on the length of the boom, workers may have to set up cable roller sheaves and rope guides at various points along the boom.

Check for proper installation and adjustment of the boom angle indicator. If necessary, boom up slightly and engage the boom hoist pawl. Then reeve the block and/or ball. Verify that the hoist wire rope is spooled tightly onto the drum and engaged with the proper sheaves. If applicable, check that the gantry wedge socket securely anchors the wire rope.

2.1.3 Raising the Boom

Manufacturers may provide a boom raising checklist to help operators verify that the crane configuration is correct before raising the boom.

Figure 15 Boom head supported by blocking.

Boom leverage on the crane's components is most extreme when raising the boom from blocking on the ground, in a horizontal position. Examples of items that may be included on such a checklist include the following:

- Crane is on a firm, level surface.
- Boom hinge pins are fully engaged and the pin locking cover is in the Up position.
- Crawler connecting pins are engaged and locking pins installed.
- Carbody jack pads are stored.
- Carbody jacks are fully retracted and pinned in stored position.
- Boom and jib inserts are installed in proper sequence per the rigging drawing.
- Spreader (if required) is installed at the proper location and spread to the proper position.
- Intermediate suspension is installed, if required (for longer booms).
- Links are properly connected between all straps.
- All insert and pendant connector pins are installed. Cotter pins are installed and spread.
- The gantry is fully raised.
- Mast arms are fully lowered.
- The mast arms bypass handle is turned off.
- Setup mode is turned off (select and confirm desired LMI operating mode).
- Boom hoist wire rope is spooled tightly onto the drum and engaged with proper sheaves. Wire rope is securely anchored to the wedge socket at the gantry.
- Load lines are spooled tightly onto the drums and engaged with proper sheaves. Load lines are securely anchored to the wedge sockets at boom and jib point or at load block and weight ball.

Figure 16 Cable reel.

- All blocking, tools, and other items have been removed from the boom and jib.
- Automatic boom stop is properly installed and adjusted.
- Cable from the crane control system is connected to the cable reel and boom butt.
- Block-up limit control is properly installed and operational; LMI is properly installed and operational.
- Crane and attachments are properly lubricated.
- Crawlers are blocked if required per load chart.
- Wind is within allowable limits for the operation given in load chart.

When ready to raise the boom to the operating position with the standard hoisting attachments, confirm that the crane's configuration agrees with the load chart for this operation. Do not lift the load block or ball when raising the boom; there is no need to place an even greater load on the crane. Pay out the hoist line to keep the additional weight on the ground. Raise the boom smoothly in one continuous lift until reaching the minimum allowed operating boom angle. Once the boom is in the air, check that the boom hoist limiting device (hoist kick-out) is installed and in working order.

> **CAUTION**
>
> Always follow the manufacturer's recommendations when initially raising the boom to avoid damaging the equipment.

2.2.0 Assembling and Raising Long Lattice Booms

The safe assembly of longer lattice booms requires some additional considerations. There is no absolute definition for a long lattice boom compared to a short one. The main differences are as follows:

- The tendency of long lattice booms to sag as they come closer to a horizontal position
- The effect of a long boom's leverage on crane stability

This section provides general guidelines for assembling longer lattice booms. Operators and A/D personnel must always follow the crane manufacturer's specific instructions.

2.2.1 Preparing for Boom Assembly

The preparations for assembling a longer boom are essentially the same as those for a shorter boom. Consult the crane's load chart and documentation for the counterweight required for raising the assembled boom and for the planned lifts. The manufacturer may require the installation of counterweights at other locations on the crane for this purpose. For very long booms, some manufacturers provide a separate wheeled counterweight that attaches like a trailer to the main crane carbody (*Figure 17*). Boom hoist cables anchored to the counterweight trailer pass through an extra lattice-boom mast, mounted behind the main boom.

Always verify that the installed counterweight is sufficient to lift the boom off the ground. If so equipped and if required, extend the extendible counterweight. Due to the larger moment forces exerted when raising long booms, some crane manufacturers recommend blocking the front idlers on the crawler ends.

Figure 17 Wheeled counterweights.

2.2.2 Boom Assembly

Install the assembly slings and/or pendants to the boom mast and assemble the base section of the boom to the upperworks. Then proceed to assemble the boom inserts in the order specified by the manufacturer. Generally, manufacturers require placing the shorter inserts closest to the base section. Assembly workers should lay out and draw together some, but not all, of the sections. The total length of the assembled boom attached beyond the boom pendants must not exceed the maximum cantilever length specified by the manufacturer (*Figure 18*). The cantilever length is the length of a horizontal boom that projects beyond the support of the boom pendants. The portion of the boom beyond the pendants is being supported at only one end, like a cantilever. The maximum cantilever length is usually a structural limitation of the boom, rather than a crane stability issue. Align the upper pin connection points and insert those pins only, securing them with pin retainers.

With the base section attached by the upper connector pins, boom up slightly until the lower pin connections align. Engage the boom hoist pawl, and install the lower pins and pin retainers (*Figure 19*).

Before moving to assemble the next boom section, workers need to extend the boom pendants to the next position along the boom, according to the documentation. Lower the boom onto blocking until the pendants are slack, then disconnect the pendants from the boom. Connect their free ends to the pendants leading to the end of the last attached insert (*Figure 20*) using pendant assembly pins and retainers. Remember to watch for boom twist. Use pendant spreader bars (*Figure 21*) as required by the manufacturer.

Assemble the next inserts in the same manner, following the manufacturer's procedure. Lower the boom onto blocking until the pendants are slack. Move the pendants out to the end of the last attached boom section (*Figure 22*). Note how the shorter inserts are near the base of the boom. Repeat for the remaining inserts (*Figure 23*).

Repeat the assembly procedure for the boom tip if not already completed. Lower the boom onto the blocks again and move the boom pendants to their operating position near the end of the boom (*Figure 24*).

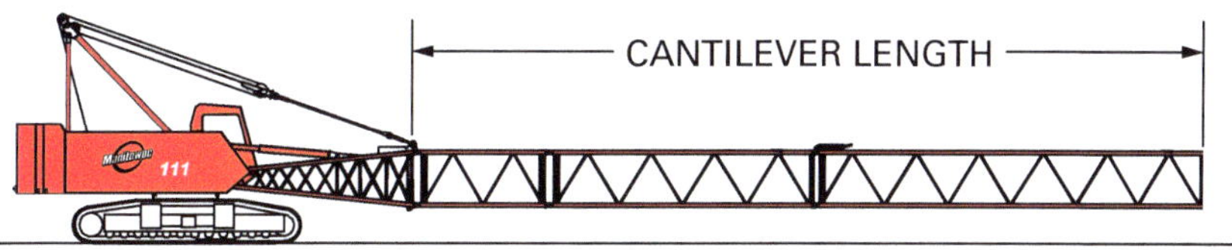

Figure 18 A cantilevered boom.

Figure 19 Securing the lower boom connector pins.

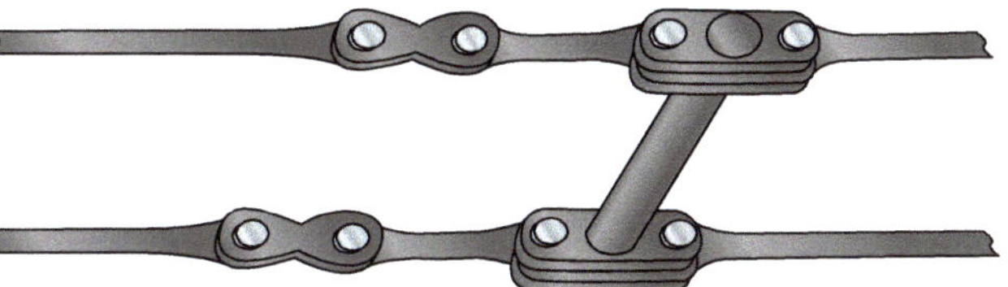

Figure 21 Pendant spreader bar.

2.2.3 Setting Up Hoist Line Components

After assembly of the lattice boom is complete, assembly workers must turn their attention to rigging the hoist line, reeving the sheaves, and attaching the hoist block or ball. Rig the main hoist line and, if necessary, the whip line along the boom. Generally, this process requires attaching or erecting boom roller sheaves and guides on the designated boom sections to support the lines between the hoist drum(s) and boom head. If necessary, boom up slightly and engage the boom hoist pawl. Reeve the boom head sheaves and block.

> **NOTE**
>
> Information regarding the installation of wire rope and reeving the load block can be found in Module 21204 of this curriculum.

2.2.4 Raising the Boom After Assembly

Some manufacturers recommend that assembly workers and operators perform certain checks before raising the boom. Refer to the checklist provided earlier for raising the boom.

> **CAUTION**
>
> When initially lifting a long lattice boom following assembly, some manufacturers require an assist crane to suspend and/or lift the boom.

> **WARNING!**
>
> If the pendants are too short on a crane with a longer boom, it will overload the boom, mast, pendants, and boom hoist systems when handling near-capacity loads.

Operators should verify that the length of the final set of pendants is the length specified in the documentation for assembly. Depending on the boom manufacturer and the electrical features provided, workers will also need to make the electrical connections along the boom between the boom head and the crane junction box on the upperworks. Verify the proper routing and securement of the electrical cable(s) to avoid pinching during boom movement. The documentation will identify the cables and the point in the procedure for completing this part of the assembly.

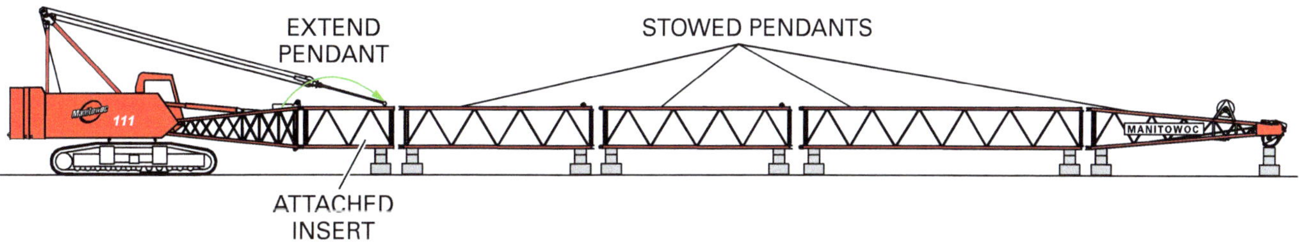

Figure 20 Move the pendants to the first boom insert.

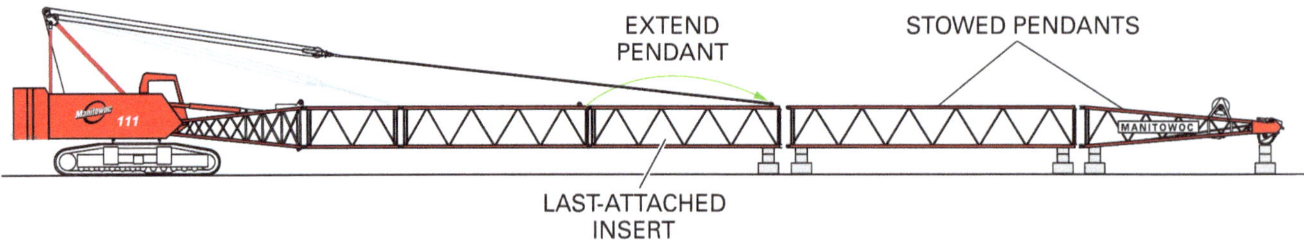

Figure 22 Extending the pendants to the last attached insert.

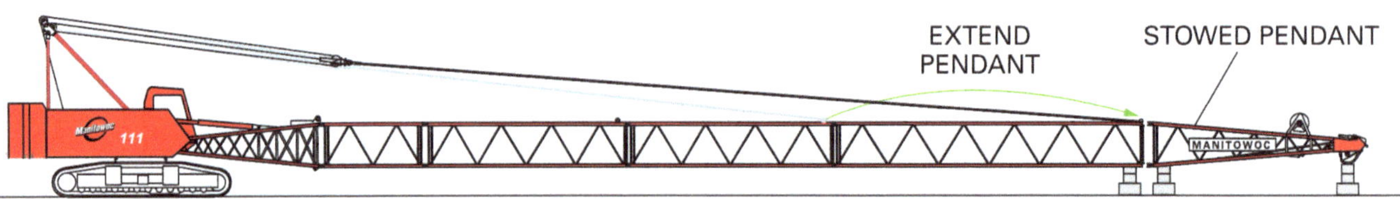

Figure 23 Completing the assembly of the boom inserts.

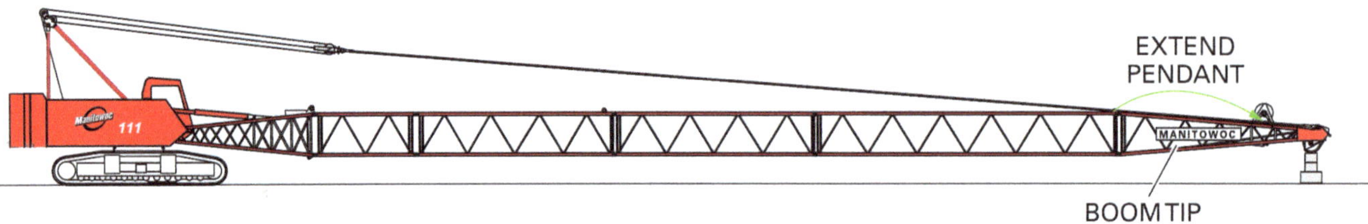

Figure 24 Extending the pendants to the boom tip.

Long lattice booms can be several hundred feet long. If the boom hoisting pendants connect to just the boom head, the boom structure between the base and head can visibly sag as it approaches horizontal. Sagging places excessive stress on the boom chords and lacing, leading to structural damage. Crane manufacturers can minimize sagging by requiring the use of boom suspension rigging, often called belly lines. These may be separate pendants attached to the boom hoisting bail and secured to the boom at a location near its midpoint. Belly lines may also be wire rope pendants reeved through sheaves at the boom hoisting bail and attached to the boom pendants, as shown in *Figure 25*. This configuration equalizes the tension between the main hoist pendants and belly lines, which can vary depending on the boom angle.

The other challenge with long booms is the excessive leverage the boom imposes on the crane when it is at or near horizontal. There are methods that can be used to solve this problem. The crane may require additional counterweight. As noted earlier, this may include counterweights on a separate carrier attached to the crane base. Some manufacturers provide for a longer live mast (*Figure 26*), which supplies the needed leverage to lift the boom. Finally, an assist crane may lift the boom tip into the permitted operating region of the crane's load chart. The manufacturer's documentation will provide directions on which method to use.

When the A/D director gives permission, raise the boom slowly. To gain extra stability when raising a boom without assistance, do not lift the load block. Pay out the hoist line to keep the block on the ground. If an assist crane is involved, coordinate with the other operator and maintain tension in your crane's boom hoist rigging as the assist crane lifts the boom. If the crane model has a manual LMI bypass feature, be sure to set the proper LMI mode when the boom is within normal operating limits. Detach the assist crane when the boom is within the operating limits of the crane.

> **WARNING!**
>
> If a worker must climb the boom to release the assist crane, that person must be trained for and use the appropriate fall arrest PPE.

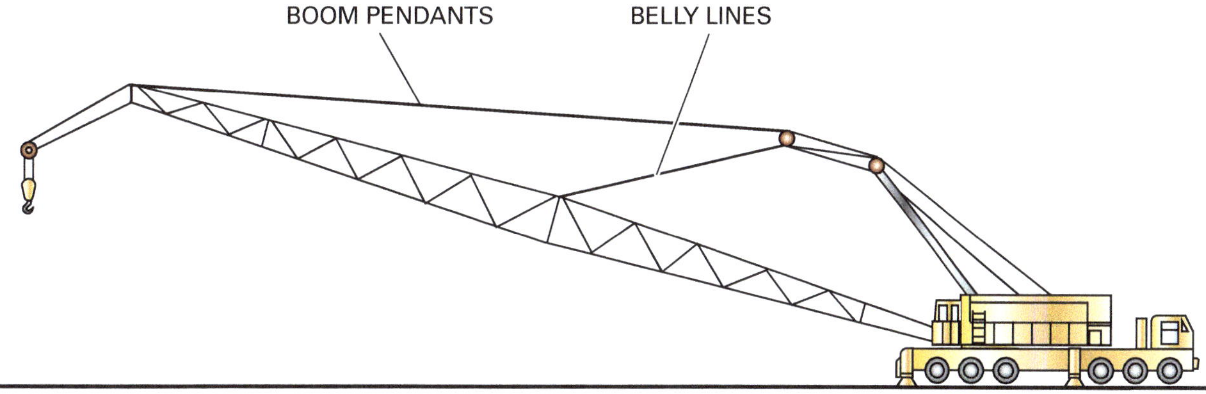

Figure 25 Belly lines suspending a long lattice boom.

Figure 26 Long live mast for operations with small boom angles.

2.3.0 Jib Assembly

After assembling the boom, workers may assemble and attach a jib extension as part of the process. The advantage of installing a jib extension during boom assembly is that the main boom is already at ground level on blocks. Adding an extension to an operating crane requires lowering the boom to ground level, which often involves significant effort.

If necessary, lower the boom onto blocking supporting the boom head, but clear of the sheaves. Pay out the main or auxiliary hoist line as required, and lay it on the ground off to the appropriate side of the boom per the documentation. Some boom tips have welded brackets to attach a jib, as shown in *Figure 27*. If a jib adapter is required for the boom, install the jib adapter to the boom tip.

Assemble the jib resting on blocking on the ground. Jibs may be a single section or they may have multiple sections. Next, workers attach the base of the jib to the boom head. They can do this by having the operator lower the main boom to

Figure 27 Boom tip with jib bracket.

the jib or having an assist crane raise the jib to the boom head. Workers then pin the jib base to the boom tip.

The process of rigging the jib's stays or operating gear depends entirely on the type of jib and whether it has remote luffing capability. A jib that can be offset manually may have nothing but the baseplate attachment points where assembly workers insert pins to establish the required offset angle. If the jib is of the fixed, A-frame type, workers must install the jib mast (*Figure 28*) first, then secure the A-frame by installing the appropriate fore- and backstay pendants for the required offset. Pin the forestay and backstays to the strut tip using the pins furnished with the pendants.

If the jib extension is a luffing jib, workers must reeve the appropriate jib hoist lines to the jib backstay. Hydraulic connections are required for a power luffing jib. The details of these assemblies are beyond the scope of this module.

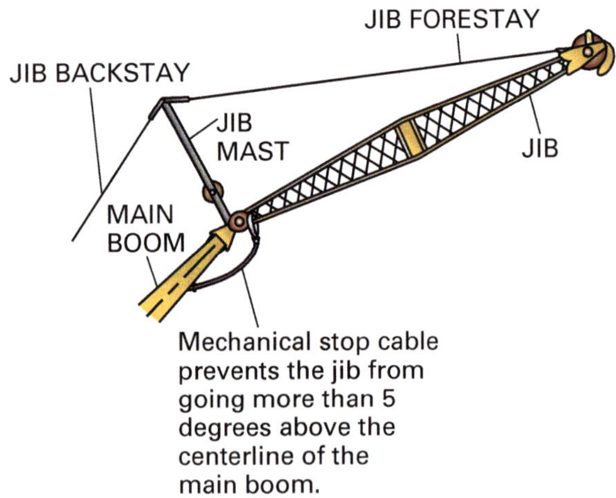

Figure 28 Jib assembly with forestay and backstay pendants.

To complete the jib assembly, erect or install a wire rope roller guide on each jib extension section used, if applicable. Reeve the hoist rope through the jib roller guide(s) and the jib head sheave. Attach the hook block or ball. Attach and connect the anti-two-block sensor to the jib hoist rope, if appropriate. Also make up any electrical connections between the jib and main boom at the main boom head. When the jib assembly is complete, raise the boom as previously described, observing the applicable precautions.

2.4.0 Disassembly of Lattice Booms

This section provides general guidelines for the disassembly of a lattice boom. Always consult the manufacturer's documentation when performing any kind of disassembly. Prepare the crane for the disassembly process, which can impose unusual stress on the crane. Remember that OSHA standards require that an A/D director be present for disassembly as well as assembly.

2.4.1 Crane Preparation

Move the crane to a broad, level, firm area with enough room to lower and land the boom. Set up barricades to minimize personnel traffic during the disassembly operations. Review the previous sections for other A/D site considerations, if necessary. After positioning the crane, fully extend the outriggers and level the base; block the idlers on a crawler, if required. Verify that the counterweight is correct, and set the swing lock.

> **CAUTION**
>
> Operators should make a thorough inspection of the boom hoist cable before booming down. Ropes damaged during crane operations can break under the increased tension imposed at minimum boom angles.

Lower the load block and/or ball to the ground, then lower the boom in the quadrant of maximum stability or as previously determined by the A/D director. For certain crane models, the operator may have to disable certain LMI safety features or switch to a different operating mode to lower the boom for disassembly. Place blocking under the boom tip and under both sides of every joint. Lower the boom onto this blocking until the pendants go slack. The process may require an assist crane to help lower long booms.

Splitting the Load

Some crane manufacturers have developed a special feature for assisting the erection of very long jibs. One or two permanently attached, lightweight wheels support the jib head on the ground. After the jib extension and its rigging are attached, the operator booms up on the main boom while paying out on the boom luffing hoist pendants. The jib head wheels roll toward the crane as the jib base rises. With this feature, the ground supports half the weight of the jib while raising the boom. At the appropriate point permitted by the load chart, the operator hauls in on the jib luffing hoists, raising the far end of the jib into position.

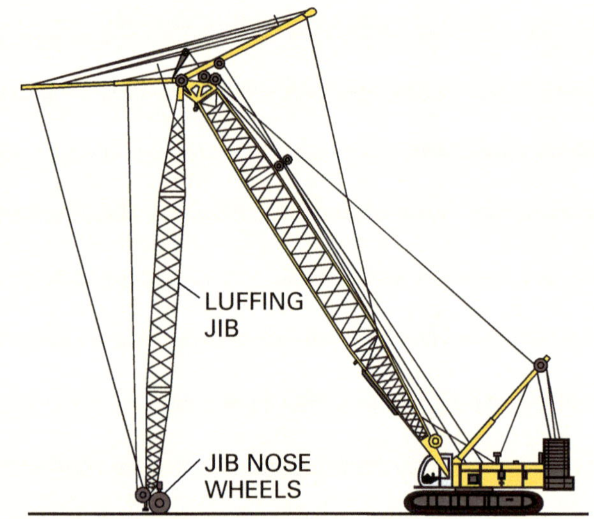

Figure Credit: Liebherr USA, Co.

Disconnect and remove the anti-two-block weights and switches. Disconnect the headache ball and unreeve the hook block according to the documentation. Pay in the load ropes, making sure they don't snag on the boom's roller sheaves. Disconnect all electrical cables along the length of the boom. If applicable, coil up the slack in electrical cables and plug their ends into the dummy socket or connectors provided for this purpose. Remove and stow any removable boom mounted equipment, such as high-voltage proximity sensors, cameras, anemometer, or FAA warning light.

2.4.2 Boom Disassembly

Disassembly of a lattice boom is simply the reverse of the assembly process described earlier. As always, refer to the specific crane manufacturer's documentation for the disassembly procedure. With the head of the boom resting on blocking and the boom pendants slack, shorten the pendants by removing a length and re-anchoring them at a lower boom section. Properly stow and secure the disconnected pendants. Observe the maximum cantilever length limitations during the disassembly process, if applicable. Elevate the boom slightly to minimize stress on the outer section's connector points, but do not elevate it so far that the boom head lifts off its blocking.

> **CAUTION**
>
> Never remove boom connector pins for sections between the pendant support points and the boom base. This will cause the boom to buckle. Always work on sections beyond the point where the pendants support the boom.

> **WARNING!**
>
> Workers must not perform boom disassembly while standing directly under or inside of a boom. Boom collapse can occur during disassembly, resulting in serious injury or death.

After positioning the boom, workers can remove the lower pins on the sections ahead of the pendants. This may require some experimenting with the boom angle to produce the slack in the connectors needed to permit removal of the pins. After properly stowing the removed pins, lower the boom back down onto the blocking. When the blocking for the outer sections bears their weight, workers can remove the upper pins on the sections ahead of the pendants. Store the removed pins in the brackets provided for this purpose (*Figure 29*) if available.

The operator may have to back the crane away to provide some working space between boom sections as the disassembly progresses. Repeat the following process until all the inserts are separated and resting on blocking:

- Disconnect and move the pendants inward to the next insert.
- Raise the boom so that workers can remove the lower connector pins.
- Lower the boom end onto blocking to remove the upper pins.

If required, workers must set up the gantry and/or mast rigging to remove the base section. Many cranes have the capability of self-disassembly. They can stage the boom sections for transport or load them onto trailers using the base section for lifting. At the appropriate point in disassembly, workers will disassemble the crane base itself, which is essentially the reverse of the process described previously.

Figure 29 Boom pin properly stowed.

Additional Resources

ASME P30.1, Planning for Load Handling Activities. Current edition. The American Society of Mechanical Engineers. Current Edition. New York, NY: American Society of Mechanical Engineers.

Mobile Crane Safety Manual. 2014. Milwaukee, WI: Association of Equipment Manufacturers.

29 *CFR* 1926, Subpart CC, *Cranes and Derricks in Construction*. **www.ecfr.gov**

2.0.0 Section Review

1. All the following actions are appropriate when preparing to assemble a lattice boom *except* _____.

   a. extending and locking outriggers to their maximum extent
   b. removing the crawler assemblies
   c. erecting the gantry and boom mast rigging
   d. extending a crane's extendable counterweight

2. The maximum cantilever length of a lattice boom is the maximum _____.

   a. length of a luffing jib
   b. load chart radius of the boom
   c. length of a lowered boom beyond the pendants
   d. boom length permitted by the manufacturer

3. When installing a jib on a lattice boom, attaching the jib may require a _____.

   a. belly line
   b. jib adapter
   c. diagonal strut
   d. back hitch assembly

4. Which boom connector pins are removed first during disassembly?

   a. Lower pins
   b. Upper pins
   c. Upper, then lower pins on the boom head section
   d. Upper, then lower pins only on the intermediate sections

SUMMARY

Lifting operations depend on the proper setup of the crane at the work site. This involves assembly of the crane base, mounting the proper counterweights, and assembly of the lattice boom. Preparation for boom assembly begins with identifying a suitable area for crane assembly.

Every crane has specific manufacturer instructions for boom assembly and disassembly. It is essential that operators and A/D personnel follow these procedures. Like all crane operations, successful crane assembly and disassembly involves planning, coordination among different groups of workers, and supervision by a qualified A/D director, as required by safety and industry standards.

1. The function of a gantry on a crane is to _____.
    a. support a jib extension at various offsets
    b. support the boom base hinge pins
    c. position the crawler assemblies during assembly
    d. provide a mechanical advantage for boom hoisting

2. When lifting lattice boom sections, it is permissible to _____.
    a. attach the hook directly to a boom main chord
    b. use bridles at boom lifting lugs
    c. use bridles attached to boom section connector pins
    d. lift by the boom lattice lacing

3. Which assembly area preparation is intended to minimize hazards to people not involved in the assembly of a crane?
    a. Checking the ground bearing pressure
    b. Posting trip hazard signs near stacks of blocking
    c. Erecting a barricade around the assembly area
    d. Providing a traffic director for trucks transporting boom components

4. Five important steps for assembling a lattice boom are listed below. Which sequence shows the typical order for these procedures?

    (A) install counterweight sections
    (B) install crawler assemblies
    (C) erect gantry and boom mast
    (D) deploy outriggers
    (E) set down on crawlers

    a. A, B, C, D, E
    b. E, D, C, B, A
    c. D, C, B, E, A
    d. A, B, E, C, D

5. Which sequence of steps describes the proper method for continuing the attachment of a boom insert to a previously assembled boom section?
    a. Move pendants out onto previous section, align connectors with next section, insert upper connector pins, boom up, and insert lower connector pins.
    b. Boom up, move pendants out onto next section, boom down, insert lower connector pins to previous section, boom up, and insert upper connector pins.
    c. Insert all lower connector pins for remaining sections, move pendants to boom head, boom up, insert all upper connector pins, boom down to set boom head on blocking.
    d. Insert all upper connector pins, move pendants to boom head, boom up, insert all lower connector pins.

6. Which of the following items refers to the LMI state when checking the crane's readiness to raise the boom following its assembly?
    a. Crawler-crane carbody jacks retracted and pinned in stowed position
    b. Spreader installed (if required) at proper location
    c. Setup mode is turned off
    d. Automatic boom stop installed

7. A long lattice boom is generally considered _____.
    a. one that is longer than 75 feet
    b. one that is longer than 150 feet
    c. any that tends to sag as it comes closer to a horizontal position
    d. one that requires FAA collision avoidance lights to be turned on at night

8. Which of the listed tasks must assembly workers complete first when attaching a jib extension?
    a. Rig the jib mast
    b. Secure the backstay pendant
    c. Reeve the fly sheave
    d. Pin the jib to the boom head

9. When installing a jib extension, one of the final steps is to _____.

   a. install the jib base pins
   b. erect the jib mast
   c. reeve the hoisting line and attach the block or ball
   d. install the forestay and backstay pendants

10. When disassembling a lattice boom, begin the process by _____.

   a. removing the lower connector pins between the boom pendant and base hinge point
   b. disconnecting the boom head and working toward the boom base
   c. disconnecting the boom base and working toward the boom head
   d. removing all the upper boom connector pins first

Trade Terms Introduced in This Module

Back hitch: The mechanism that permits adjusting the position of the boom gantry to maximize its leverage on the boom hoist cables.

Belly lines: Additional pendants attached to the boom hoisting bail and secured at a location near the midpoint of a long boom to minimize sagging.

Boom stops: Devices that mechanically prevent the boom from exceeding a maximum boom angle, reducing the potential for backward tipping or structural damage.

Cantilever length: The length of a horizontal boom that projects beyond the support of the boom pendants, leaving that portion of the boom supported at only one end, like a cantilever.

Diagonal struts: In a lattice boom with a rectangular cross section, pieces of structural pipe or angle iron welded to diagonally opposite chords of the boom to stiffen the assembly.

Gantry: A heavy frame with roller sheaves mounted on the upperworks of a crane, used for providing a mechanical advantage to the boom hoist ropes; also refers to a jib strut or mast.

Hoist kick-out: A term for any safety device that automatically stops a hoist drum when a limit is reached.

Lattice lacing: In a lattice boom, the zig-zag pattern of structural metal supporting the main chords along the sides, top, and bottom of the boom sections.

Live boom mast: A heavy framework or lattice boom with sheaves, hinged at the lower end of its base on the crane's upperworks, whose purpose is to control the angle of the main boom through attached boom pendants. Boom hoist cables reeved through the boom gantry sheaves or hydraulic actuators control its position.

Main chords: The heavy structural pipe or angle iron members running the length of a boom section.

Pendant: A length of heavy wire rope or flexibly connected sections of flat metal bars, equipped with connectors at both ends, used to hoist and/or support sections of a crane's boom.

Suspension rope: The rope reeved through a crane's gantry and mast sheaves to the sheaves suspending the boom pendants.

MANITOWOC MODEL 18000 PRODUCT GUIDE – BOOM COMBINATIONS

boom combinations

No. 55 or No. 55A Boom Combinations

Boom Length m (ft)	Boom Inserts		
	3,0 m (10 ft)	6,1 m (20 ft)	12,2 m (40 ft)
36,6 (120)	1	1	1
42,7 (140)	1	–	2
48,8 (160)	1	1	2
54,9 (180)	1	2	2
61,0 (200)	1	1	3
67,1 (220)	1	2	3
73,2 (240)	1	1	4
79,2 (260)	1	2	4
85,3 (280)	1	1	5
91,4 (300)	1	2	5
97,5 (320)	1	1	6

NOTE: 36,6 m (120') basic boom consists of 6,1 m (20') butt, 3,0 m (10') insert with drum, 6,1 m (20') insert, 12,2 m (40'), and 9,1 m (30') top.

May use 12,2 m (40') No. 55 insert with or without sheaves.

No. 55-79A Boom Combinations

Boom Length m (ft)	Boom Inserts					No. 55-79A Transition
	55*			79		
	3,0 m (10 ft)	6,1 m (20 ft)	12,2 m (40 ft)	6,1 m (20 ft)	12,2 m (40 ft)	12,2 m (40 ft)
36,6 (120)	1	1	–	–	–	1
42,7 (140)	1	–	1	–	–	1
48,8 (160)	1	1	1	–	–	1
54,9 (180)	1	1	1	1	–	1
61,0 (200)	1	1	1	–	1	1
67,1 (220)	1	1	1	1	1	1
73,2 (240)	1	1	1	–	2	1
79,2 (260)	1	1	1	1	2	1
85,3 (280)	1	1	1	–	3	1
91,4 (300)	1	1	1	1	3	1
97,5 (320)	1	1	1	–	4	1
103,6 (340)	1	1	1	1	4	1
109,7 (360)	1	1	1	–	5	1
115,8 (380)	1	1	1	1	5	1
121,9 (400)	1	1	1	–	6	1

NOTE: 36,6 m (120') basic boom consists of 6,1 m (20') butt, 3,0 m (10') insert with drum, 6,1 m (20') insert, 12,2 m (40'), and 9,1 m (30') top.

May use 12,2 m (40') No. 55 insert with or without sheaves.

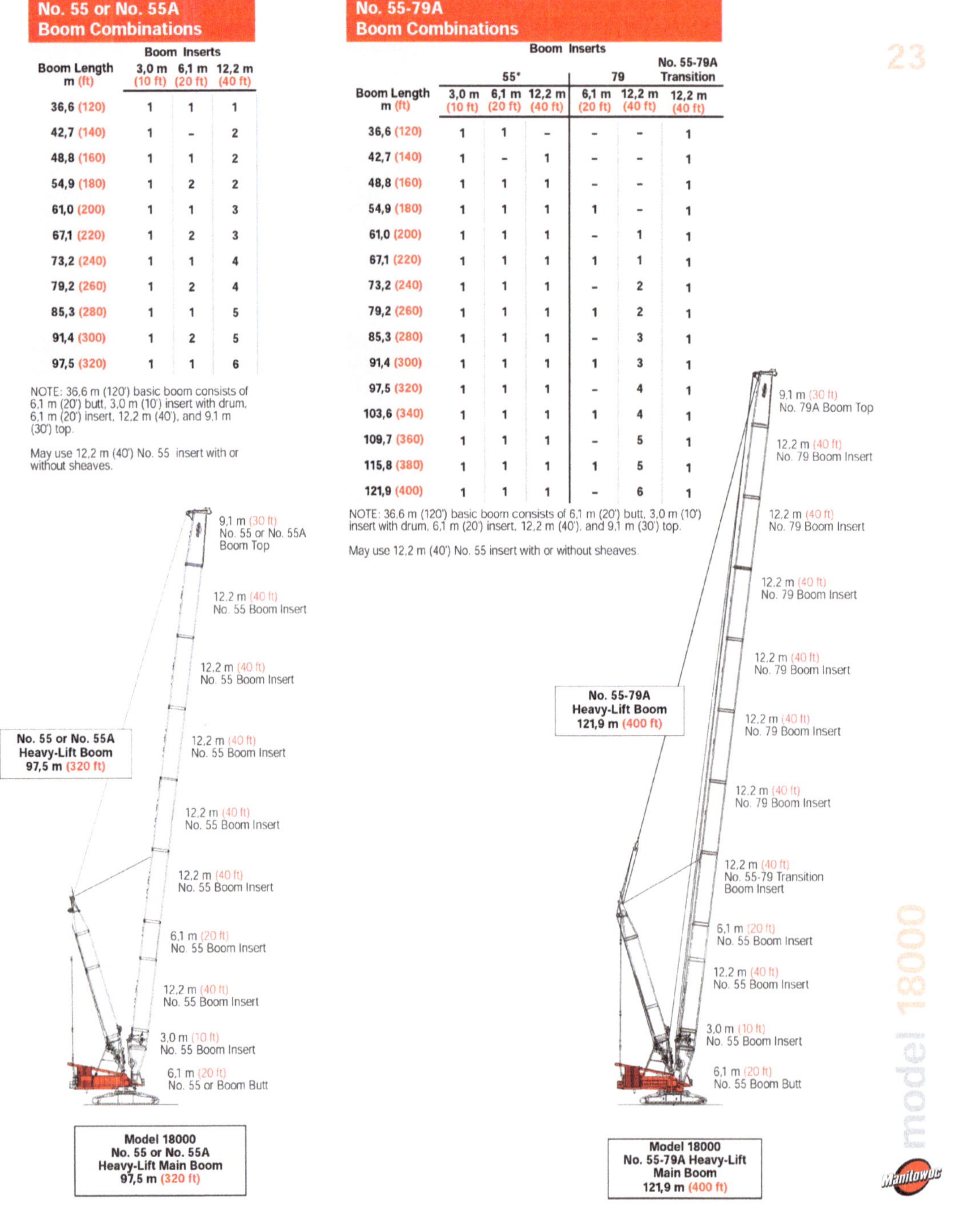

boom combinations

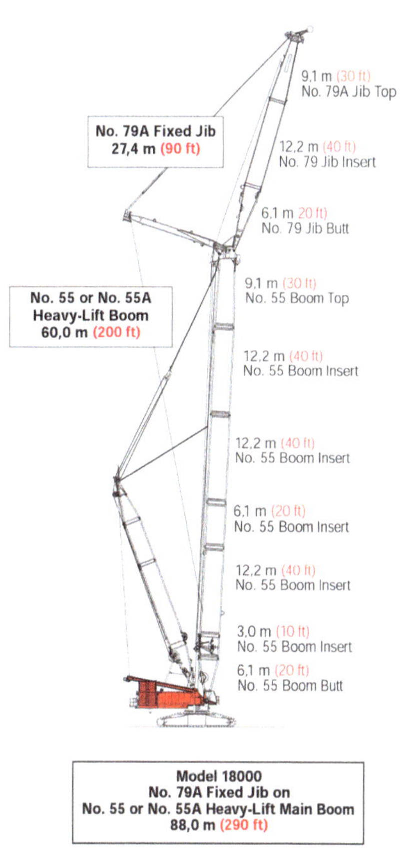

boom combinations

No. 44 Luffing Jib Combinations

Luffing Jib Length m (ft)	Boom Inserts		
	3,0 m (10 ft)	6,1 m (20 ft)	12,2 m (40 ft)
21,3 (70)	–	–	–
24,4 (80)	1	–	–
27,4 (90)	–	1	–
30,5 (100)	1	1	–
33,5 (110)	–	–	1
36,6 (120)	1	–	1
39,6 (130)	–	1	1
42,7 (140)	1	1	1
45,7 (150)	–	–	2
48,8 (160)	1	–	2
51,8 (170)	–	1	2
54,9 (180)	1	1	2
57,9 (190)	–	–	3
61,0 (200)	1	–	3
64,0 (210)	–	1	3
67,1 (220)	1	1	3
70,1 (230)	–	–	4
73,2 (240)	1	–	4

No. 79A Luffing Jib Combinations

Luffing Jib Length m (ft)	Boom Inserts	
	6,1 m (20 ft)	12,2 m (40 ft)
27,4 (90)	2	–
33,5 (110)	1	1
39,6 (130)	2	1
45,7 (150)	1	2
51,8 (170)	2	2
57,9 (190)	1	3
64,0 (210)	2	3
70,1 (230)	1	4
76,2 (250)	2	4
82,3 (270)	1	5
88,3 (290)	2	5
94,5 (310)	1	6

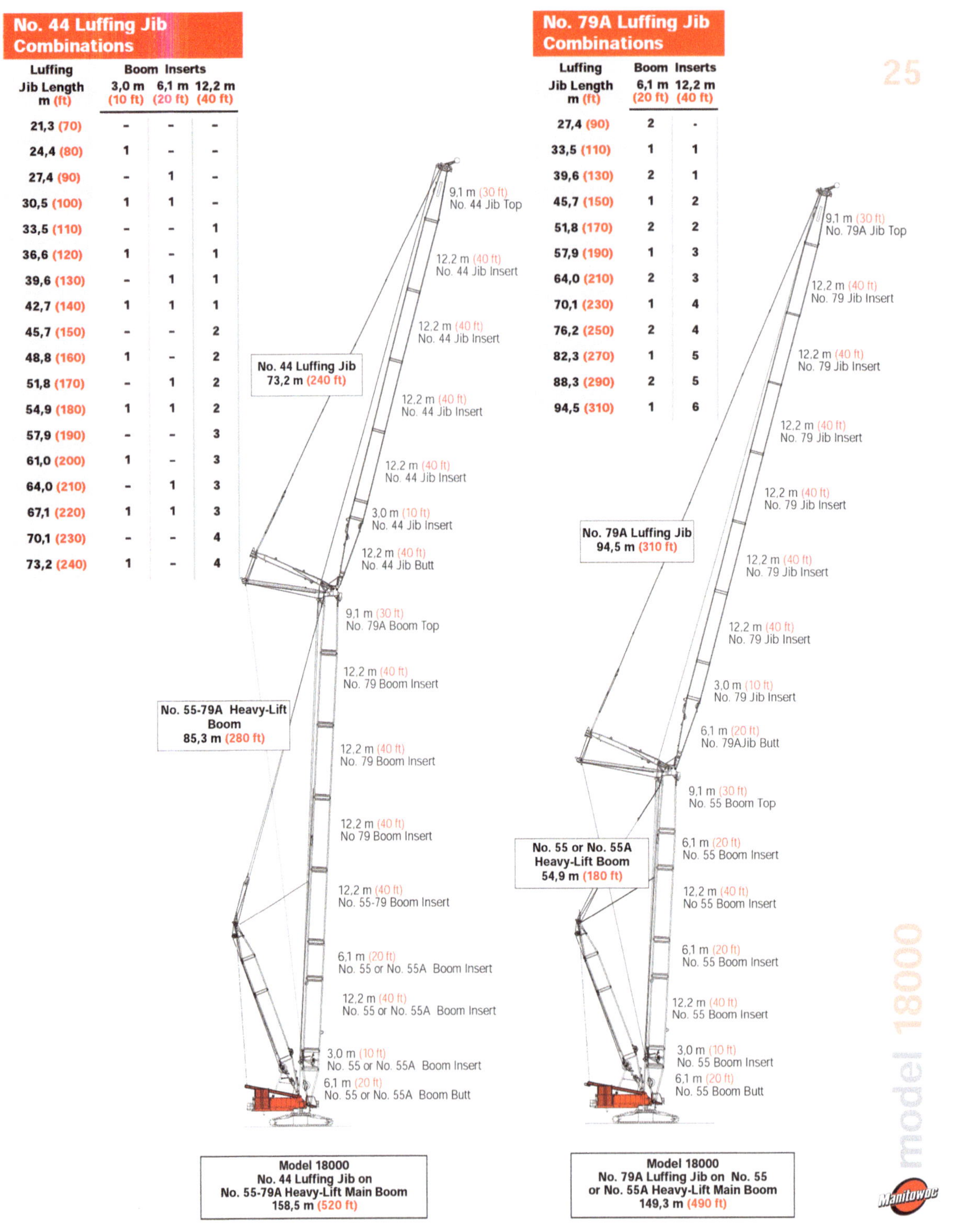

Model 18000
No. 44 Luffing Jib on No. 55-79A Heavy-Lift Main Boom
158,5 m (520 ft)

Model 18000
No. 79A Luffing Jib on No. 55 or No. 55A Heavy-Lift Main Boom
149,3 m (490 ft)

Additional Resources

This module is intended as a thorough resource for task training. The following reference materials are recommended for further study.

ASME P30.1, Planning for Load Handling Activities. Current edition. The American Society of Mechanical Engineers. Current Edition. New York, NY: American Society of Mechanical Engineers.

Mobile Crane Safety Manual. 2014. Milwaukee, WI: Association of Equipment Manufacturers.

29 CFR 1926, Subpart CC, *Cranes and Derricks in Construction*. **www.ecfr.gov**

Figure Credits

Link-Belt Construction Equipment Company, Module opener, Figures 10, 27

The Manitowoc Company, Inc., Figures 1, 5, 11–14 (line art), 18, Appendix

Mammoet USA South Inc., Figure 9, 26

Liebherr USA, Co., SA01

Section Review Answer Key

Answer	Section Reference	Objective
Section One		
1. d	1.0.0	1a
2. b	1.2.0	1b
Section Two		
1. b	2.1.1	2a
2. c	2.2.2	2b
3. b	2.3.0	2c
4. a	2.4.2	2d

NCCER – *Intermediate Rigger*

NCCER CURRICULA — USER UPDATE

NCCER makes every effort to keep its textbooks up-to-date and free of technical errors. We appreciate your help in this process. If you find an error, a typographical mistake, or an inaccuracy in NCCER's curricula, please fill out this form (or a photocopy), or complete the online form at **www.nccer.org/olf**. Be sure to include the exact module ID number, page number, a detailed description, and your recommended correction. Your input will be brought to the attention of the Authoring Team. Thank you for your assistance.

Instructors – If you have an idea for improving this textbook, or have found that additional materials were necessary to teach this module effectively, please let us know so that we may present your suggestions to the Authoring Team.

NCCER Product Development and Revision

13614 Progress Blvd., Alachua, FL 32615

Email: curriculum@nccer.org
Online: www.nccer.org/olf

❏ Trainee Guide   ❏ Lesson Plans   ❏ Exam   ❏ PowerPoints   Other _______________________

Craft / Level: _______________________________________ Copyright Date: _______________

Module ID Number / Title: ____

Section Number(s): ____

Description: ____

____

____

____

Recommended Correction: ____

____

____

____

Your Name: ____

Address: ____

____

Email: ____ Phone: _________________

Glossary

Acceleration: In common usage, the rate of increase of an object's speed per second. The rate of decrease of speed is commonly called *deceleration*.

Anchorage point: A point used to fix the position of a block or other type of hoist or pulling device. An anchorage point must be able to easily handle the load weights and line pulls associated with any sling angle imposed on it, without fear of failure.

Back hitch: The mechanism that permits adjusting the position of the boom gantry to maximize its leverage on the boom hoist cables.

Belly lines: Additional pendants attached to the boom hoisting bail and secured at a location near the midpoint of a long boom to minimize sagging.

Birdcaging: A term used to describe the condition where wire rope is forced into compression along its length, pushing the outer strands away from the core and/or inner strands, forming a shape that resembles a bird cage.

Boom stops: Devices that mechanically prevent the boom from exceeding a maximum boom angle, reducing the potential for backward tipping or structural damage.

Bridle slings: Lift rigging consisting of multi-leg, wire rope or chain slings.

Buoyancy: The upward force exerted by a liquid or gas (both fluids) on an object immersed in the fluid.

Cantilever length: The length of a horizontal boom that projects beyond the support of the boom pendants, leaving that portion of the boom supported at only one end, like a cantilever.

Center of gravity (CG): The point at which the entire weight of an object is concentrated, such that supporting the object above this specific point would result in its remaining balanced in position.

Center of gravity (CG): The point where an object's mass, and therefore its weight, is concentrated. The concept is useful for determining stability, the balance point, and leverage.

Center of rotation: In crane operations, the vertical axis through the center of the swing circle around which the upperworks rotates.

Choke angle: The outside angle between the vertical part of a choker hitch and the part of the sling passing through the eye, hook, or shackle forming the choke.

Coefficient of friction (CF): A ratio that expresses a comparison between the force necessary to move an object over the surface of another material, and the pressure between the two materials.

Constructional stretch: The stretching that occurs in wire rope when it is initially placed in service, during its break-in period. Constructional stretch is expected in all wire ropes. Constructional stretch in standard ropes is generally 0.25 to 1 percent of the rope's length.

Core: The axial part at the center of wire rope around which the individual strands are laid.

Crane moment: The product of the entire crane's center of gravity and its distance from the tipping axis along the lever arm toward the load's center of gravity. Commonly called the crane's *leverage* or *lever force*.

Crossover points: Per 29 *CFR* 1926.1401, specific locations where a wire rope layer spooled on a drum must climb up and cross over the previous layer. Crossover points are located at each end of the drum at its flanges, where the rope must change direction to continue spooling onto the drum.

D/d ratio: For rope-like slings used in hitches that wrap around the load, the ratio of the diameter of the load to the diameter of the rope. Manufacturers establish minimum D/d ratios to avoid overstressing the sling with a bend that is too sharp.

D/d ratio: The ratio of a sheave's pitch diameter, represented by D, and the body diameter of the wire rope, represented by d, that is passing over it.

Dead end: The excess wire rope exposed after a termination has been properly fitted. This part of the wire rope is not exposed to the strain applied by a load to the wire rope.

Derate: The process of reducing the working load limit of a sling according to standard rules to account for the large stresses that exist in a sling when used in certain applications.

Design factor: The ratio of a wire rope's minimum breaking strength to the maximum load expected to be applied.

Diagonal struts: In a lattice boom with a rectangular cross section, pieces of structural pipe or angle iron welded to diagonally opposite chords of the boom to stiffen the assembly.

Drifting: The process of moving a load laterally through the air, using multiple hoists and transferring the load weight from one hoist to another.

Dynamics: A general term that refers to forces, their effects, and how they change from one moment to the next.

Effective weight: Also referred to as the *apparent weight*. The weight of a load plus or minus the effect of other factors, such as acceleration, wind, and buoyancy. For example, a load that accelerates while falling exerts a force that is greater than the actual weight of the load when it is suddenly stopped.

Fatigue fractures: Progressive fractures resulting from the repeated bending of individual wires. These fractures may occur at bending-stress levels well below the ultimate strength of the material. Lack of lubrication and rust on a wire rope can accelerate fatigue fractures.

Fiber core (FC): A cord or rope of natural or synthetic fiber used as the axial center (core) of a wire rope.

Filler wire: Small wire used to create spacers within a strand that help hold the position and support the other strand wires.

Flange points: Per 29 *CFR* 1926.1401, points where the wire rope contacts the drum flanges at each end.

Flemish eyes: Spliced loops formed in the end of wire rope rigging slings. For personnel hoist rigging, the eyes must also contain a thimble and be mechanically crimped with a ferrule.

Fulcrum: The point of support on which a lever pivots.

Gantry: A heavy frame with roller sheaves mounted on the upperworks of a crane, used for providing a mechanical advantage to the boom hoist ropes; also refers to a jib strut or mast.

Hitch: Any method of attaching a sling to a load. Common hitches include vertical, bridle, basket, and choker.

Hoist kick-out: A term for any safety device that automatically stops a hoist drum when a limit is reached.

Impact loading: A short-duration force that is suddenly exerted when a moving object is slowed or stopped.

Independent wire rope core (IWRC): A wire rope used as the axial center (core) of a larger wire rope assembly.

Invert: To move a load from the horizontal position, for example, to any other position, such as over on its side, vertical on its end, or fully inverted to lie on its top.

Lattice lacing: In a lattice boom, the zig-zag pattern of structural metal supporting the main chords along the sides, top, and bottom of the boom sections.

Law of moments: A scientific principle that defines the conditions under which a lever-and-fulcrum system is balanced and stable.

Lay length: The distance measured parallel to the axis of the rope or strand in which a strand or wire makes one complete revolution around the core or center.

Lay: In the context of wire rope, the lay describes the direction of rotation of the strands in a wire rope assembly.

Lever: A simple machine consisting of a rigid arm pivoting on a fulcrum. Force can be transferred from one end to the other as it pivots, changing the force's direction and size, or two forces acting on its ends can work against each other.

Leverage: A common term describing the action of a lever, especially for prying or lifting a heavy object. See *moment*.

Live boom mast: A heavy framework or lattice boom with sheaves, hinged at the lower end of its base on the crane's upperworks, whose purpose is to control the angle of the main boom through attached boom pendants. Boom hoist cables reeved through the boom gantry sheaves or hydraulic actuators control its position.

Live section: Any part of a wire rope that is exposed to the load weight.

Load moment indicator (LMI): A system that aids the equipment operator by directly or indirectly sensing the overturning moment on the equipment. It compares the lifting conditions to the equipment's rated capacity, and when the rated capacity is reached, it disables equipment functions that can increase the severity of loading on the equipment, such as hoisting or telescoping the boom further out.

Load moment: The force applied to the crane by the load; the leverage of the load, opposing the leverage of the crane. The load moment is calculated by multiplying the gross load weight by the horizontal distance from the tipping fulcrum to the center of gravity of the suspended load. The load moment is usually reported to the operator as a percentage of the crane's capacity at the present set of conditions. As those conditions change, such as the boom angle, the load moment changes as well.

Load radius: The horizontal distance between the crane's center of rotation and the vertical hoist line or tackle with the load applied. On load charts, the load radius is determined by the boom angle and its length to the boom tip.

Main chords: The heavy structural pipe or angle iron members running the length of a boom section.

Minimum breaking strength: Also referred to as *minimum breaking force*. The documented breaking strength of a wire rope product established under set industry testing procedures.

Moment arm: In a lever system, the distance between the point where a force is applied to a lever and the lever's fulcrum.

Moment: The effect of a force acting perpendicularly on a lever at a certain distance from its fulcrum. Commonly called *leverage*. Mathematically, moment is equal to the product of the force and its moment arm.

Momentum: A property of all moving objects that is equal to the product of their mass and velocity. In equation form, $M = m \times v = mv$, where M is the object's momentum, m is its mass, and v is its velocity.

Monel™: A trademarked metal alloy made primarily from nickel and copper, with small amounts of other materials added. Monel is resistant to corrosion initiated by a number of common sources, including seawater, and is an expensive material.

Nondestructive testing (NDT): Testing the integrity of fabricated metal structures, especially welds, by various means to detect hidden flaws without destroying the object or rendering it useless. Common NDT methods include dye penetrant (PT), magnetic particle (MT), X-ray and radiography (RT), and ultrasound (UT).

Pendant: A length of heavy wire rope or flexibly connected sections of flat metal bars, equipped with connectors at both ends, used to hoist and/or support sections of a crane's boom.

Pendants: Fixed lengths of rope with mechanical fittings at each end.

Pendulum effect: The variable force exerted by a swinging load on the crane structure along the hoist line, like the swing of a pendulum. The load's excess force is maximum at the bottom of the swing.

Pitch diameter: The diameter of the sheave measured at its root where the bottom of the wire rope makes contact as it travels over. The pitch diameter represents the size of the circle formed by the bottom of the rope if it traveled 360° around the sheave.

Poured socket: Referring to wire rope terminations, a method of attaching a fitting by placing the end of the rope inside the fitting socket, then pouring molten zinc, another metal alloy, or a special resin into the socket. After hardening, these terminations can provide 100 percent of the wire rope's minimum breaking strength. This process is also referred to as *speltering*.

Preformed strand: A strand constructed of wires that were formed into the helical shape they would assume once placed in the strand, before the strand was assembled.

Preformed wire rope: A wire rope constructed of strands that were formed into the helical shape they would assume once placed in the rope, before the rope was assembled.

Proof test: A strength test of an object, usually under a load exceeding that expected under normal use, to ensure it is suitable for safe use. In personnel hoisting, a proof test requires loading the platform with a weight equal to 125 percent of its rated load and suspending it for at least 5 minutes.

Repetitive pickup points: Per 29 *CFR* 1926.1401, refers to the sections of a rope where short-cycle operations cause it to be repeatedly spooled on and off a small portion of the drum.

Rotation-resistant wire rope: Wire rope constructed with inner and outer strands placed in opposing lays, rather than using the same lay in both layers. This causes the torque of the strand layers to oppose each other, significantly reducing the tendency to twist.

Running ropes: Wire ropes that move over sheaves and/or drums.

Seizing: The securing of an open end of wire rope by wrapping wire tightly around its circumference several times. This action prevents fraying and unraveling of the wire rope. This preparation is used on both sides of a planned wire rope cut.

Sling angle: The angle formed between the sling and the horizontal when under tension.

Sling-angle factor (SAF): A factor that permits calculating the tension in a sling supporting a known load at an angle. Equal to the length of the sling between its bearing points divided by the vertical distance between those points (SAF = L ÷ H).

Speltered: To be attached with molten zinc or a zinc alloy. The term *spelter* is considered a synonym for *zinc*.

Stability: A measure of a movable object's ability to resist a change in position of its center of gravity. The more force required to raise its CG, the more stable it is.

Strand patterns: The various configurations of the wires used to form a strand. The four basic strand patterns are wire, Seale, Warrington, and filler wire.

Strands: A single grouping of round or shaped wires laid helically around a core or center wire. A wire rope assembly for lifting purposes is typically a collection of strands around a common core.

Suspension rope: The rope reeved through a crane's gantry and mast sheaves to the sheaves suspending the boom pendants.

Swaged: A connection made by forming metal, usually with a die or shaped tool, to change the diameter. A swaged wire rope connection is made by reducing the diameter of the socket around the rope.

Tipping axis: In crane operations, an imaginary horizontal line around which a crane's center of gravity would rotate as it tips under the right circumstances.

Trial lift: For a personnel hoisting platform, a lift of the platform and rigging without occupants, but with an equivalent load, under the same conditions and over the same route as the lift with personnel will take.

Tuggers: Motorized or engine-powered winches designed to replace manpower to pull the rope used for lifting or moving equipment.

Wire strand core (WSC): A wire strand assembly used as the axial center (core) member of a wire rope.

Wire: A single, continuous length of metal, usually cold-drawn or cold-rolled from a rod, available in various diameters.

Working load limit (WLL): The weight capacity a manufacturer certifies a rigging component can lift under stated conditions. The standard sling WLL is provided for a vertical hitch. Riggers will typically apply standard rules to derate WLLs when slings are used in other configurations.